SBAC Math Practice Grade 5

Complete Content Review Plus 2 Full-length SBAC Math Tests

Elise Baniam - Michael Smith

SBAC Math Practice Grade 5
Published in the United State of America By
The Math Notion
Email: info@Mathnotion.com
Web: WWW.MathNotion.com

Copyright © 2020 by the Math Notion. All rights reserved. No part of this publication may be reproduced, stored in a retrieval system, or transmitted in any form or by any means, electronic, mechanical, photocopying, recording, scanning, or otherwise, except as permitted under Section 107 or 108 of the 1976 United States Copyright Ac, without permission of the author.
All inquiries should be addressed to the Math Notion.

ISBN: 978-1-63620-021-7

About the Author

Elise Baniam has been a math instructor for over a decade now. She graduated in Mathematics. Since 2006, Elise has devoted his time to both teaching and developing exceptional math learning materials. As a Math instructor and test prep expert, Elise has worked with thousands of students. She has used the feedback of her students to develop a unique study program that can be used by students to drastically improve their math score fast and effectively.

– **SAT Math Workbook**
– **ACT Math Workbook**
– **ISEE Math Workbooks**
– **SSAT Math Workbooks**
–**many Math Education Workbooks**
– **and some Mathematics books ...**

As an experienced Math teacher, Mrs. Baniam employs a variety of formats to help students achieve their goals: she teaches students in large groups, and she provides training materials and textbooks through her website and through Amazon.

You can contact Elise via email at:
Elise@Mathnotion.com

Get the Targeted Practice You Need to Excel on the Math Section of the SBAC Test Grade 5!

SBAC Math Practice Book Grade 5 is **an excellent investment in your future** and the best solution for students who want to maximize their score and minimize study time. Practice is an essential part of preparing for a test and improving a test taker's chance of success. The best way to practice taking a test is by going through lots of SBAC math questions.

High-quality mathematics instruction ensures that students become problem solvers. We believe all students can develop deep conceptual understanding and procedural fluency in mathematics. In doing so, through this math workbook we help our students grapple with real problems, think mathematically, and create solutions.

SBAC Math Practice Book allows you to:

- Reinforce your strengths and improve your weaknesses
- Practice **2500+ realistic** SBAC math practice questions
- Exercise math problems in a variety of formats that provide intensive practice
- Review and study **Two Full-length SBAC Practice Tests** with detailed explanations

...and much more!

This Comprehensive SBAC Math Practice Book is carefully designed to provide only that **clear and concise information** you need.

WWW.MathNotion.com

... So Much More Online!

✓ FREE Math Lessons

✓ More Math Learning Books!

✓ Mathematics Worksheets

✓ Online Math Tutors

For a PDF Version of This Book

Please Visit WWW.MathNotion.com

Contents

Chapter 1: Place Value and Number Sense 11
   Numbers in Standard Form 12
   Number in Expand Form 13
   Odd or Even 14
   Compare Whole Numbers 15
   Pattern 16
   Round Whole Numbers 17
   Answer key Chapter 1 18

Chapter 2: Whole Number Operations 21
   Order of Operations 22
   Estimate Sums 23
   Estimate Differences 24
   Subtract from Whole Thousands. 25
   Multiplication Whole Number 26
   Long Division by Two Digit 27
   Division with Remainders 27
   Dividing Hundreds 28
   Answer key Chapter 2 29

Chapter 3: Number Theory 31
   Factoring 32
   Prime Factorization 33
   Divisibility Rule 34
   Great Common Factor (GCF) 35
   Least Common Multiple (LCM) 36
   Answer key Chapter 3 37

Chapter 4: Fractions 39
   Adding Fractions – Like Denominator 40
   Adding Fractions – Unlike Denominator 41
   Subtracting Fractions – Like Denominator 42
   Subtracting Fractions – Unlike Denominator 43
   Converting Mix Numbers 44

Converting improper Fractions 45
Adding Mix Numbers 46
Subtracting Mix Numbers 47
Simplify Fractions... 48
Multiplying Fractions.. 49
Multiplying Mixed Number 50
Dividing Fractions 51
Dividing Mixed Number 52
Comparing Fractions 53
Answer key Chapter 4.. 54

Chapter 5: Decimal 59
Graph Decimals.. 60
Round Decimals 61
Decimals Addition 62
Decimals Subtraction 63
Decimals Multiplication 64
Decimal Division... 65
Comparing Decimals 66
Convert Fraction to Decimal 67
Answer key Chapter 5.. 68

Chapter 6: Exponent and Radicals... 71
Positive Exponents 72
Negative Exponents 73
Add and subtract Exponents 74
Exponent multiplication 75
Exponent division... 76
Scientific Notation 77
Square Roots 78
Simplify Square Roots 79
Answer key Chapter 6.. 80

Chapter 7: Ratio, Proportion and Percent... 83
Proportions... 84
Reduce Ratio 85
Word Problems 86

Percent 87
Convert Fraction to Percent 88
Convert Decimal to Percent 89
Answer key Chapter 7 90

Chapter 8: Measurement 93
Reference Measurement 94
Metric Length Measurement 95
Customary Length Measurement 95
Metric Capacity Measurement 96
Customary Capacity Measurement 96
Metric Weight and Mass Measurement 97
Customary Weight and Mass Measurement 97
Time 98
Answers of Worksheets – Chapter 8 99

Chapter 9: Algebraic Expressions 101
Find a Rule 102
Variables and Expressions 103
Translate Phrases 104
Distributive Property 105
Evaluate One Variable Expressions 106
Answer key Chapter 9 107

Chapter 10: Symmetry and Transformations 109
Line Segments 110
Parallel, Perpendicular and Intersecting Lines 111
Identify Lines of Symmetry 112
Lines of Symmetry 113
Identify Three–Dimensional Figures 114
Vertices, Edges, and Faces 115
Identify Faces of Three–Dimensional Figures 116
Answers of Worksheets – Chapter 10 117

Chapter 11: Geometry 121
Area and Perimeter of Square 122
Area and Perimeter of Rectangle 123
Area and Perimeter of Triangle 124

Area and Perimeter of Trapezoid 125
Area and Perimeter of Parallelogram 126
Circumference and Area of Circle 127
Perimeter of Polygon 128
Volume of Cubes 129
Volume of Rectangle Prism 130
Answer key Chapter 11 131

Chapter 12: Data and Graphs 133
Mean and Median 134
Mode and Range 135
Stem-And-Leaf Plot 136
Dot plots 137
Bar Graph 138
Probability 139
Answer key Chapter 12 140

SBAC Test Review 143
SBAC Practice Test 1 147
SBAC Practice Test 2 159

Answers and Explanations 171
Answer Key 173
Practice Test 1 175
Practice Test 2 179

Chapter 1:
Place Value and Number Sense

Numbers in Standard Form

Write the number in standard form.

1) 10 million 208 thousand 24

    _____

2) 72 million 9 thousand 708

    _____

3) 121 million 24 thousand 453

    _____

4) 541 million 75 thousand 127

    _____

5) 90 billion 15 million 68 thousand 15

    _____

6) 12 billion 120 million 5

    _____

7) 8 billion 114 million 88 thousand

    _____

8) 16 billion 28 thousand 785

    _____

9) 75 billion 159 thousand 324

    _____

10) 41 billion 3 million 8 thousand 25

    _____

11) 16 billion 129 thousand 989

    _____

12) 65 billion 220 million 6 thousand 2

    _____

13) 785 million 124 thousand 97

    _____

14) 33 billion 104 million 11 thousand 57

    _____

15) 95 billion 424 million

    _____

16) 27 billion 77 million 9 thousand 150

    _____

Number in Expand Form

Write the number in expand form.

1) 956: _____.

2) 3,800: _____.

3) 52,457: _____.

4) 60,070: _____.

5) 409,389: _____.

6) 76,805: _____.

7) 745,321: _____.

8) 8,146: _____.

9) 19,037: _____.

10) 52,799: _____.

11) 5,125: _____.

12) 400,544: _____.

13) 600,700: _____.

14) 3,080,000: _____.

Odd or Even

Write odd or even.

1) 19 _____

2) 81 _____

3) 456 _____

4) 852 _____

5) 953 _____

6) 183 _____

7) 987 _____

8) 540 _____

9) 777 _____

10) 544 _____

11) 33 _____

12) 4,458 _____

13) 15,159 _____

14) 9,357 _____

15) 3,000 _____

16) 14 _____

17) 257 _____

18) 660 _____

19) 45,789 _____

20) 15,300 _____

21) 452 _____

22) 49,459 _____

23) 84 _____

24) 7,700 _____

25) 6,451 _____

26) 985 _____

Compare Whole Numbers

Compare, writing <, >, or = between the numbers.

1) 40,420 ☐ 41,004

2) 29,460 ☐ 29,640

3) 78,920 ☐ 87,290

4) 34,570 ☐ 33,750

5) 96,328 ☐ 96,238

6) 85,843 ☐ 85,840

7) 76,584 ☐ 76,854

8) 72,998 ☐ 72,989

9) 37,467 ☐ 37,567

10) 48,878 ☐ 49,878

11) 56,660 ☐ 65,660

12) 73,898 ☐ 69,899

13) 89,990 ☐ 98,110

14) 84,760 ☐ 84,670

15) 26,680 ☐ 26,860

16) 86,440 ☐ 86,440

17) 158,980 ☐ 158,890

18) 201,807 ☐ 201,807

19) 243,240 ☐ 243,420

20) 345,566 ☐ 354,655

21) 187,158 ☐ 196,001

22) 137,983 ☐ 137,895

23) 278,788 ☐ 249,988

24) 194,854 ☐ 194,845

25) 219,390 ☐ 291,110

26) 305,288 ☐ 299,999

27) 317,857 ☐ 371,857

28) 405,710 ☐ 405,170

Pattern

Continue this pattern for four more numbers:

1) 1,400; 1,250; 1,100; 950; _____

2) 3,700; 3,500; 3,300; 3,100; _____

3) 4,200; 3,850; 3,500; 3,150; _____

4) 1,650; 1,530; 1,410; 1,290; _____

5) 2,900; 2,650; 2,400; 2,150; _____

6) 5,000; 4,600; 4,200; 3,800; _____

7) 3,950; 3,700; 3,450; 3,200; _____

8) 2,800; 2,675; 2,550; 2,425; _____

9) 3,850; 3,550; 3,250; 2,950; _____

10) 4,700; 4,500; 4,300; 4,100; _____

11) Write a list of five numbers that follows this pattern: Start at 250 and add 200 each time.

_____

Round Whole Numbers

Round to the place of the underlined digit.

1) 7,467,589 ≈ _____

2) 546,125 ≈ _____

3) 9,187,208 ≈ _____

4) 15,685,807 ≈ _____

5) 5,454,676 ≈ _____

6) 3,588,975 ≈ _____

7) 8,368,519 ≈ _____

8) 27,754,769 ≈ _____

9) 42,654,411 ≈ _____

10) 7,621,879 ≈ _____

11) 19,788,987 ≈ _____

12) 4,286,850 ≈ _____

13) 9,273,778 ≈ _____

14) 6,484,684 ≈ _____

15) 5,157,628 ≈ _____

16) 8,667,885 ≈ _____

17) 3,567,980 ≈ _____

18) 8,369,432 ≈ _____

19) 24,256,880 ≈ _____

20) 5,229,758 ≈ _____

21) 6,987,422 ≈ _____

22) 4,877,391 ≈ _____

Answer key Chapter 1

Numbers in Standard Form

1) 10,208,024
2) 72,009,708
3) 121,024,453
4) 541,075,127
9) 75,000,159,324
10) 41,003,008,025
11) 16,000,129,989
12) 65,220,006,002
5) 90,015,068,015
6) 12,120,000,005
7) 8,114,088,000
8) 16,000,028,785
13) 785,124,097
14) 33,104,011,057
15) 95,424,000,000
16) 27,077,009,150

Numbers in Expand Form

1) $(9 \times 100) + (5 \times 10) + 6$
2) $(3 \times 1,000) + (8 \times 100)$
3) $(5 \times 10,000) + (2 \times 1,000) + (4 \times 100) + (5 \times 10) + 7$
4) $(6 \times 10,000) + (7 \times 10) + 0$
5) $(4 \times 100,000) + (9 \times 1,000) + (3 \times 100) + (8 \times 10) + 9$
6) $(7 \times 10,000) + (6 \times 1,000) + (8 \times 100) + 5$
7) $(7 \times 100,000) + (4 \times 10,000) + (5 \times 1,000) + (3 \times 100) + (2 \times 10) + 1$
8) $(8 \times 1,000) + (1 \times 100) + (4 \times 10) + 6$
9) $(1 \times 10,000) + (9 \times 1,000) + (3 \times 10) + 7$
10) $(5 \times 10,000) + (2 \times 1,000) + (7 \times 100) + (9 \times 10) + 9$
11) $(5 \times 1,000) + (1 \times 100) + (2 \times 10) + 5$
12) $(4 \times 100,000) + (5 \times 100) + (4 \times 10) + 4$
13) $(6 \times 100,000) + (7 \times 100)$
14) $(3 \times 1,000,000) + (8 \times 10,000)$

Odd or Even

1) Odd
2) Odd
3) Even
4) Even
5) Odd
6) Odd
7) Odd
8) Even
9) Odd
10) Even
11) Odd
12) Even

13) Odd
14) Odd
15) Even
16) Even
17) Odd
18) Even
19) Odd
20) Even
21) Even
22) Odd
23) Even
24) Even
25) Odd
26) Odd

Compare Whole Numbers

1) <
2) <
3) <
4) >
5) >
6) >
7) <
8) >
9) <
10) <
11) <
12) >
13) <
14) >
15) <
16) =
17) >
18) =
19) <
20) <
21) <
22) >
23) >
24) >
25) <
26) >
27) <
28) >

Pattern

1) 800; 650; 500; 350
2) 2,900; 2,700; 2,500; 2,300
3) 2,800; 2,450; 2,100; 1,750
4) 1,170; 1,050; 930; 810
5) 1,900; 1,650; 1,400; 1,150
6) 3,400; 3,000; 2,600; 2,200
7) 2,950; 2,700; 2,450; 2,200
8) 2,300; 2,175; 2,050; 1,925
9) 2,650; 2,350; 2,050; 1,750
10) 3,900; 3,700; 3,500; 3,300
11) 250; 450; 650; 850; 1,050

Round whole number

1) 7,470,000
2) 546,000
3) 9,187,000
4) 15,686,000
5) 5,455,000
6) 3,589,000
7) 8,368,500
8) 27,754,800
9) 42,654,400
10) 7,600,000
11) 19,789,000
12) 4,287,000
13) 9,270,000
14) 6,485,000
15) 5,157,600
16) 8,668,000
17) 3,568,000
18) 8,369,430
19) 24,257,000
20) 5,230,000
21) 6,990,000
22) 4,877,000

Chapter 2: Whole Number Operations

Order of Operations

Evaluate each expression.

1) $4 + (3 \times 2) =$

2) $19 - (5 \times 3) =$

3) $(12 \times 4) + 11 =$

4) $(19 - 4) - (2 \times 6) =$

5) $45 + (24 \div 6) =$

6) $(15 \times 3) \div 9 =$

7) $(81 \div 9) \times (-2) =$

8) $(6 \times 9) + (28 - 19) =$

9) $70 + (4 \times 5) + 2 =$

10) $(25 \times 2) \div (5 + 5) =$

11) $(-8) + (11 \times 4) + 12 =$

12) $(5 \times 8) - (40 \div 8) =$

13) $(9 \times 4 \div 3) - (15 + 8) =$

14) $(12 + 5 - 10) \times 7 - 10 =$

15) $(48 - 28 + 10) \times (88 \div 11) =$

16) $32 + (18 - (27 \div 9)) =$

17) $(8 + 17 - 3 - 6) + (36 \div 4) =$

18) $(75 - 25) + (18 - 14 + 6) =$

19) $(30 \times 2) + (15 \times 3) - 12 =$

20) $19 + 17 - (24 \times 4) + 13 =$

Estimate Sums

Estimate the sum by rounding each added to the nearest ten.

1) $29 + 8 =$

2) $38 + 37 =$

3) $26 + 19 =$

4) $47 + 28 =$

5) $11 + 39 =$

6) $18 + 33 =$

7) $58 + 21 =$

8) $46 + 68 =$

9) $58 + 78 =$

10) $51 + 72 =$

11) $56 + 29 =$

12) $56 + 13 =$

13) $46 + 49 =$

14) $28 + 75 =$

15) $76 + 93 =$

16) $36 + 49 =$

17) $56 + 72 =$

18) $12 + 47 =$

19) $93 + 27 =$

20) $15 + 83 =$

21) $34 + 74 =$

22) $43 + 73 =$

23) $124 + 282 =$

24) $68 + 64 =$

Estimate Differences

Estimate the difference by rounding each number to the nearest ten.

1) 49 − 12 =

2) 62 − 52 =

3) 86 − 57 =

4) 61 − 43 =

5) 77 − 58 =

6) 39 − 17 =

7) 56 − 29 =

8) 78 − 61 =

9) 89 − 46 =

10) 91 − 69 =

11) 98 − 63 =

12) 63 − 17 =

13) 94 − 58 =

14) 69 − 48 =

15) 82 − 52 =

16) 46 − 22 =

17) 68 − 51 =

18) 91 − 66 =

19) 78 − 67 =

20) 96 − 38 =

21) 71 − 64 =

22) 87 − 31 =

23) 96 − 28 =

24) 86 − 43 =

Subtract from Whole Thousands.

Find the sum or difference.

1) 1,000 − 9 = ____

2) 3,000 − 7 = ____

3) 8,000 − 5 = ____

4) 6,000 − 20 = ____

5) 9,000 − 10 = ____

6) 5,000 − 35 = ____

7) 7,000 − 70 = ____

8) 4,000 − 15 = ____

9) 3,000 − 9 = ____

10) 6,000 − 40 = ____

11) 9,000 − 300 = ____

12) 8,000 − 6 = ____

13) 7,000 − 25 = ____

14) 5,000 − 150 = ____

15) 8,000 − 450 = ____

16) 4,000 − 40 = ____

17) 6,000 − 110 = ____

18) 9,000 − 150 = ____

19) 5,000 − 600 = ____

20) 3,000 − 18 = ____

21) 7,000 − 140 = ____

22) 3,000 − 55 = ____

23) 2,000 − 50 = ____

24) 4,000 − 12 = ____

25) 6,000 − 35 = ____

26) 9,000 − 80 = ____

27) 5,000 − 40 = ____

28) 3,000 − 90 = ____

Multiplication Whole Number

Calculate.

1) $170 \times 8 =$

2) $145 \times 30 =$

3) $(-9) \times 7 \times (-6) =$

4) $-6 \times (-8) \times (-8) =$

5) $14 \times (-14) =$

6) $50 \times (-6) =$

7) $9 \times (-3) \times 7 =$

8) $(-300) \times (-20) =$

9) $(-40) \times (-40) \times 2 =$

10) $150 \times 4 =$

11) $135 \times 30 =$

12) $312 \div 12 =$

13) $(-3,300) \div 4 =$

14) $(-36) \div (-9) =$

15) $272 \div (-17) =$

16) $3,000 \div 24 =$

17) $(-168) \div 4 =$

18) $3,096 \div 3 =$

19) $1,044 \div (-29) =$

20) $5,880 \div 7 =$

21) $(-81) \div 3 =$

22) $(-4,000) \div (-40) =$

23) $0 \div 450 =$

24) $(-1,360) \div 8 =$

25) $6,408 \div 712 =$

26) $(-3,150) \div 5 =$

27) $7,268 \div 2 =$

28) $(-108) \div (-9) =$

SBAC Math Practice Grade 5

Long Division by Two Digit

Find the quotient.

1) 14)448

2) 13)884

3) 21)588

4) 22)286

5) 42)714

6) 28)252

7) 19)589

8) 37)1,887

9) 55)1,100

10) 38)1,368

11) 56)2,632

12) 60)2,880

13) 32)2,912

14) 58)6,322

15) 87)3,828

16) 82)6,642

17) 60)11,040

18) 15)9,240

Division with Remainders

Find the quotient with remainder.

1) 14)715

2) 16)2,750

3) 19)3,243

4) 79)3,478

5) 29)5,340

6) 64)6,843

7) 46)5,861

8) 78)10,340

9) 54)8,802

10) 91)13,018

11) 43)11,419

12) 55)33,872

13) 81)6,538

14) 89)35,092

Dividing Hundreds

Find answers.

1) $3,000 \div 300$

2) $3,200 \div 40$

3) $1,000 \div 200$

4) $3,600 \div 900$

5) $4,200 \div 700$

6) $1,200 \div 400$

7) $2,700 \div 900$

8) $3,500 \div 700$

9) $10,200 \div 300$

10) $20,000 \div 400$

11) $5,400 \div 200$

12) $9,600 \div 600$

13) $1,800 \div 300$

14) $7,700 \div 700$

15) $9,600 \div 800$

16) $4,500 \div 500$

17) $21,000 \div 700$

18) $6,600 \div 200$

19) $8,100 \div 100$

20) $15,000 \div 200$

21) $18,000 \div 6,000$

22) $28,000 \div 70$

23) $490 \div 70$

24) $640 \div 80$

Answer key Chapter 2

Order of Operations

1) 10	6) 5	11) 48	16) 47
2) 4	7) −18	12) 35	17) 25
3) 59	8) 63	13) −11	18) 60
4) 3	9) 92	14) 39	19) 93
5) 49	10) 5	15) 240	20) −47

Estimate sums

1) 40	7) 80	13) 100	19) 120
2) 80	8) 120	14) 110	20) 100
3) 50	9) 140	15) 170	21) 100
4) 80	10) 120	16) 90	22) 110
5) 50	11) 90	17) 130	23) 400
6) 50	12) 70	18) 60	24) 130

Estimate differences

1) 40	7) 30	13) 30	19) 10
2) 10	8) 20	14) 20	20) 60
3) 30	9) 40	15) 30	21) 10
4) 20	10) 20	16) 30	22) 60
5) 20	11) 40	17) 20	23) 70
6) 20	12) 40	18) 20	24) 50

Subtract from Whole Thousands

1) 991	10) 5,960	19) 4,400
2) 2,993	11) 8,700	20) 2,982
3) 7,995	12) 7,994	21) 6,860
4) 5,980	13) 6,975	22) 2,945
5) 8,990	14) 4,850	23) 1,950
6) 4,965	15) 7,550	24) 3,988
7) 6,930	16) 3,960	25) 5,965
8) 3,985	17) 5,890	26) 8,920
9) 2,991	18) 8,850	27) 4,960

28) 2,910

Multiplication Whole Number

1) 1,360
2) 4,350
3) 378
4) −576
5) −196
6) −300
7) −189
8) 6,000
9) 3,200
10) 600
11) 4,050
12) 26
13) −825
14) 4
15) −16
16) 125
17) −42
18) 1,032
19) −36
20) 840
21) −27
22) 100
23) 0
24) −170
25) 9
26) −630
27) 3,634
28) 12

Long Division by Two Digit

1) 32
2) 68
3) 28
4) 13
5) 17
6) 9
7) 31
8) 51
9) 20
10) 36
11) 47
12) 48
13) 91
14) 109
15) 44
16) 81
17) 184
18) 616

Division with Remainders

1) 51 R1
2) 171 R14
3) 170 R13
4) 44 R2
5) 184 R4
6) 106 R59
7) 127 R19
8) 132 R44
9) 163 R0
10) 143 R5
11) 265 R24
12) 615 R47
13) 80 R58
14) 394 R26

Dividing Hundreds

1) 10
2) 80
3) 5
4) 4
5) 6
6) 3
7) 3
8) 5
9) 34
10) 50
11) 27
12) 16
13) 6
14) 11
15) 12
16) 9
17) 30
18) 33
19) 81
20) 75
21) 3
22) 400
23) 7
24) 8

Chapter 3: Number Theory

Factoring

Factor, write prime if prime.

1) 15

2) 72

3) 25

4) 42

5) 32

6) 66

7) 34

8) 20

9) 50

10) 35

11) 40

12) 30

13) 49

14) 54

15) 96

16) 108

17) 76

18) 90

19) 100

20) 85

21) 63

22) 24

23) 48

24) 115

25) 51

26) 93

27) 52

28) 105

Prime Factorization

Factor the following numbers to their prime factors.

1.
   16
   / \
   _____

2.
   38
   / \
   _____

3.
   51
   / \
   _____

4.
   12
   / \
   _____

5.
   18
   / \
   _____

6.
   23
   / \
   _____

7.
   46
   / \
   _____

8.
   58
   / \
   _____

9.
   64
   / \
   _____

10.
   82
   / \
   _____

11.
   87
   / \
   _____

12.
   98
   / \
   _____

Divisibility Rule

Apply the divisibility rules to find the factors of each number.

1) 20   2, 3, 4, 5, 6, 9, 10   13) 24   2, 3, 4, 5, 6, 9, 10

2) 84   2, 3, 4, 5, 6, 9, 10   14) 395   2, 3, 4, 5, 6, 9, 10

3) 252   2, 3, 4, 5, 6, 9, 10   15) 920   2, 3, 4, 5, 6, 9, 10

4) 64   2, 3, 4, 5, 6, 9, 10   16) 137   2, 3, 4, 5, 6, 9, 10

5) 220   2, 3, 4, 5, 6, 9, 10   17) 440   2, 3, 4, 5, 6, 9, 10

6) 465   2, 3, 4, 5, 6, 9, 10   18) 360   2, 3, 4, 5, 6, 9, 10

7) 75   2, 3, 4, 5, 6, 9, 10   19) 495   2, 3, 4, 5, 6, 9, 10

8) 120   2, 3, 4, 5, 6, 9, 10   20) 4,870   2, 3, 4, 5, 6, 9, 10

9) 1,125   2, 3, 4, 5, 6, 9, 10   21) 590   2, 3, 4, 5, 6, 9, 10

10) 88   2, 3, 4, 5, 6, 9, 10   22) 326   2, 3, 4, 5, 6, 9, 10

11) 454   2, 3, 4, 5, 6, 9, 10   23) 114   2, 3, 4, 5, 6, 9, 10

12) 155   2, 3, 4, 5, 6, 9, 10   24) 470   2, 3, 4, 5, 6, 9, 10

Great Common Factor (GCF)

Find the GCF of the numbers.

1) 8, 26

2) 16, 44

3) 28, 38

4) 10, 35

5) 18, 48

6) 36, 52

7) 40, 75

8) 50, 45

9) 52, 8

10) 55, 85

11) 74, 94

12) 65, 20

13) 90, 10

14) 12, 34

15) 58, 86

16) 40, 95

17) 14, 56

18) 70, 100, 30

19) 64, 108

20) 63, 91

21) 20, 15, 35

22) 6, 12, 36

23) 25, 35, 70

24) 41, 39

Least Common Multiple (LCM)

Find the LCM of each.

1) 6, 14

2) 12, 18

3) 4, 16, 12

4) 10, 8

5) 10, 2, 15

6) 35, 7

7) 14, 35, 21

8) 8, 7

9) 11, 22, 44

10) 42, 21

11) 24, 72

12) 100, 25

13) 10, 5, 20

14) 15, 60

15) 20, 4, 3

16) 21, 14

17) 34, 17

18) 16, 64

19) 20, 70

20) 9, 39

21) 13, 8

22) 7, 20

23) 45, 63

24) 27, 4

Answer key Chapter 3

Factoring

1) 1, 3, 5, 15
2) 1, 2, 3, 4, 6, 8, 9, 12, 18, 24, 36, 72
3) 1, 5, 25
4) 1, 2, 3, 6, 7, 14, 21, 42
5) 1, 2, 4, 8, 16, 32
6) 1, 2, 3, 6, 11, 22, 33, 66
7) 1, 2, 17, 34
8) 1, 2, 4, 5, 10, 20
9) 1, 2, 5, 10, 25, 50
10) 1, 5, 7, 35
11) 1, 2, 4, 5, 8, 10, 20, 40
12) 1, 2, 3, 5, 6, 10, 15, 30
13) 1, 7, 49
14) 1, 2, 3, 6, 9, 18, 27, 54
15) 1, 2, 3, 4, 6, 8, 12, 16, 24, 32, 48, 96
16) 1, 2, 3, 4, 6, 9, 12, 18, 27, 36, 54, 108
17) 1, 2, 4, 19, 38, 76
18) 1, 2, 3, 5, 6, 9, 10, 15, 18, 30, 45, 90
19) 1, 2, 4, 5, 10, 20, 25, 50, 100
20) 1, 5, 17, 85
21) 1, 3, 7, 9, 21, 63
22) 1, 2, 3, 4, 6, 8, 12, 24
23) 1, 2, 3, 4, 6, 8, 12, 16, 24, 48
24) 1, 5, 23, 115
25) 1, 3, 17, 51
26) 1, 3, 31, 93
27) 1, 2, 4, 13, 26, 52
28) 1, 3, 5, 7, 15, 21, 35, 105

Prime Factorization

1) $2 \times 2 \times 2 \times 2$
2) 2×19
3) 3×17
4) $2 \times 2 \times 3$
5) $2 \times 3 \times 3$
6) 23 is a prime number
7) 2×23
8) 2×29
9) $2 \times 2 \times 2 \times 2 \times 2 \times 2$
10) 2×41
11) 3×29
12) $2 \times 7 \times 7$

Divisibility Rule

1) 20   2, 3, <u>4</u>, <u>5</u>, 6, 9, 10
2) 84   <u>2</u>, <u>3</u>, <u>4</u>, 5, <u>6</u>, 9, 10
3) 252   <u>2</u>, <u>3</u>, <u>4</u>, 5, <u>6</u>, <u>9</u>, 10
4) 64   <u>2</u>, 3, <u>4</u>, 5, 6, 9, 10
5) 220   <u>2</u>, 3, <u>4</u>, <u>5</u>, 6, 9, <u>10</u>
6) 465   2, <u>3</u>, 4, <u>5</u>, 6, 9, 10
7) 75   2, <u>3</u>, 4, <u>5</u>, 6, 9, 10
8) 120   <u>2</u>, <u>3</u>, <u>4</u>, <u>5</u>, <u>6</u>, 9, <u>10</u>
9) 1,125   2, <u>3</u>, 4, <u>5</u>, 6, <u>9</u>, 10
10) 88   <u>2</u>, 3, <u>4</u>, 5, 6, 9, 10
11) 454   <u>2</u>, 3, 4, 5, 6, 9, 10
12) 155   2, 3, 4, <u>5</u>, 6, 9, 10

SBAC Math Practice Grade 5

13) 24   <u>2</u>, <u>3</u>, <u>4</u>, 5, <u>6</u>, 9, 10

14) 395   2, 3, 4, <u>5</u>, 6, 9, 10

15) 920   <u>2</u>, 3, <u>4</u>, <u>5</u>, 6, 9, <u>10</u>

16) 137   2, 3, 4, 5, 6, 9, 10

17) 440   <u>2</u>, 3, <u>4</u>, <u>5</u>, 6, 9, <u>10</u>

18) 360   <u>2</u>, <u>3</u>, <u>4</u>, <u>5</u>, <u>6</u>, <u>9</u>, <u>10</u>

19) 495   2, <u>3</u>, 4, <u>5</u>, 6, <u>9</u>, 10

20) 4,870   <u>2</u>, 3, 4, <u>5</u>, 6, 9, <u>10</u>

21) 590   <u>2</u>, 3, 4, <u>5</u>, 6, 9, <u>10</u>

22) 326   <u>2</u>, 3, 4, 5, 6, 9, 10

23) 114   <u>2</u>, <u>3</u>, 4, 5, <u>6</u>, 9, 10

24) 470   <u>2</u>, 3, 4, <u>5</u>, 6, 9, <u>10</u>

Great Common Factor (GCF)

1) 2
2) 4
3) 2
4) 5
5) 6
6) 4
7) 5
8) 5
9) 4
10) 5
11) 2
12) 5
13) 10
14) 2
15) 2
16) 5
17) 14
18) 10
19) 4
20) 7
21) 5
22) 6
23) 5
24) 1

Least Common Multiple (LCM)

1) 42
2) 36
3) 48
4) 40
5) 30
6) 35
7) 210
8) 56
9) 44
10) 42
11) 72
12) 100
13) 20
14) 60
15) 60
16) 42
17) 34
18) 64
19) 140
20) 117
21) 104
22) 140
23) 315
24) 108

Chapter 4:
Fractions

Adding Fractions – Like Denominator

Find each sum.

1) $\dfrac{1}{3} + \dfrac{1}{3} =$

2) $\dfrac{3}{7} + \dfrac{1}{7} =$

3) $\dfrac{2}{9} + \dfrac{5}{9} =$

4) $\dfrac{7}{15} + \dfrac{1}{15} =$

5) $\dfrac{5}{23} + \dfrac{4}{23} =$

6) $\dfrac{8}{29} + \dfrac{7}{29} =$

7) $\dfrac{7}{19} + \dfrac{1}{19} =$

8) $\dfrac{5}{16} + \dfrac{1}{16} =$

9) $\dfrac{5}{31} + \dfrac{8}{31} =$

10) $\dfrac{5}{51} + \dfrac{8}{51} =$

11) $\dfrac{1}{17} + \dfrac{3}{17} =$

12) $\dfrac{2}{13} + \dfrac{4}{13} =$

13) $\dfrac{5}{41} + \dfrac{19}{41} =$

14) $\dfrac{2}{55} + \dfrac{9}{55} =$

15) $\dfrac{5}{21} + \dfrac{8}{21} =$

16) $\dfrac{10}{33} + \dfrac{2}{33} =$

17) $\dfrac{3}{11} + \dfrac{3}{11} =$

18) $\dfrac{24}{67} + \dfrac{1}{67} =$

19) $\dfrac{3}{19} + \dfrac{6}{19} =$

20) $\dfrac{20}{47} + \dfrac{13}{47} =$

Adding Fractions – Unlike Denominator

Add the fractions and simplify the answers.

1) $\frac{1}{2} + \frac{1}{6} =$

2) $\frac{2}{3} + \frac{3}{4} =$

3) $\frac{3}{5} + \frac{1}{2} =$

4) $\frac{7}{10} + \frac{1}{3} =$

5) $\frac{4}{15} + \frac{1}{5} =$

6) $\frac{3}{14} + \frac{2}{7} =$

7) $\frac{2}{7} + \frac{1}{3} =$

8) $\frac{1}{20} + \frac{1}{5} =$

9) $\frac{7}{12} + \frac{1}{6} =$

10) $\frac{1}{8} + \frac{3}{4} =$

11) $\frac{4}{21} + \frac{1}{3} =$

12) $\frac{5}{36} + \frac{2}{9} =$

13) $\frac{4}{35} + \frac{3}{7} =$

14) $\frac{2}{33} + \frac{1}{11} =$

15) $\frac{10}{27} + \frac{1}{3} =$

16) $\frac{7}{45} + \frac{4}{9} =$

17)

18) $\frac{1}{7} + \frac{2}{5} =$

19) $\frac{1}{2} + \frac{5}{22} =$

20) $\frac{3}{16} + \frac{1}{4} =$

21) $\frac{5}{16} + \frac{1}{24} =$

22) $\frac{2}{15} + \frac{3}{10} =$

23) $\frac{5}{42} + \frac{4}{21} =$

24) $\frac{1}{36} + \frac{5}{24} =$

Subtracting Fractions – Like Denominator

Find the difference.

1) $\dfrac{6}{5} - \dfrac{1}{5} =$

2) $\dfrac{7}{12} - \dfrac{4}{12} =$

3) $\dfrac{8}{13} - \dfrac{5}{13} =$

4) $\dfrac{19}{6} - \dfrac{7}{6} =$

5) $\dfrac{11}{21} - \dfrac{9}{21} =$

6) $\dfrac{15}{43} - \dfrac{7}{43} =$

7) $\dfrac{9}{23} - \dfrac{3}{23} =$

8) $\dfrac{18}{47} - \dfrac{15}{47} =$

9) $\dfrac{8}{20} - \dfrac{4}{20} =$

10) $\dfrac{34}{48} - \dfrac{17}{48} =$

11) $\dfrac{6}{7} - \dfrac{2}{7} =$

12) $\dfrac{36}{51} - \dfrac{28}{51} =$

13) $\dfrac{9}{11} - \dfrac{5}{11} =$

14) $\dfrac{29}{49} - \dfrac{14}{49} =$

15) $\dfrac{12}{17} - \dfrac{6}{17} =$

16) $\dfrac{17}{23} - \dfrac{11}{23} =$

17) $\dfrac{5}{7} - \dfrac{1}{7} =$

18) $\dfrac{12}{31} - \dfrac{8}{31} =$

19) $\dfrac{38}{61} - \dfrac{29}{61} =$

20) $\dfrac{31}{52} - \dfrac{20}{52} =$

21) $\dfrac{51}{63} - \dfrac{46}{63} =$

22) $\dfrac{55}{83} - \dfrac{25}{83} =$

23) $\dfrac{54}{77} - \dfrac{30}{77} =$

24) $\dfrac{49}{55} - \dfrac{37}{55} =$

Subtracting Fractions – Unlike Denominator

Solve each problem.

1) $\dfrac{1}{3} - \dfrac{1}{6} =$

2) $\dfrac{3}{5} - \dfrac{1}{9} =$

3) $\dfrac{1}{4} - \dfrac{1}{6} =$

4) $\dfrac{7}{9} - \dfrac{3}{10} =$

5) $\dfrac{11}{12} - \dfrac{5}{24} =$

6) $\dfrac{9}{16} - \dfrac{3}{20} =$

7) $\dfrac{19}{30} - \dfrac{1}{6} =$

8) $\dfrac{1}{2} - \dfrac{7}{15} =$

9) $\dfrac{3}{5} - \dfrac{2}{7} =$

10) $\dfrac{8}{9} - \dfrac{4}{11} =$

11) $\dfrac{6}{8} - \dfrac{7}{48} =$

12) $\dfrac{3}{4} - \dfrac{7}{10} =$

13) $\dfrac{4}{5} - \dfrac{8}{45} =$

14) $\dfrac{6}{7} - \dfrac{3}{10} =$

15) $\dfrac{11}{12} - \dfrac{13}{24} =$

16) $\dfrac{4}{9} - \dfrac{15}{63} =$

17) $\dfrac{5}{12} - \dfrac{5}{16} =$

18) $\dfrac{5}{8} - \dfrac{3}{10} =$

19) $\dfrac{5}{7} - \dfrac{3}{8} =$

20) $\dfrac{3}{4} - \dfrac{21}{44} =$

Converting Mix Numbers

Convert the following mixed numbers into improper fractions.

1) $2\frac{3}{5} =$

2) $3\frac{4}{5} =$

3) $5\frac{3}{4} =$

4) $3\frac{5}{8} =$

5) $6\frac{2}{5} =$

6) $7\frac{8}{9} =$

7) $2\frac{2}{15} =$

8) $3\frac{1}{13} =$

9) $2\frac{1}{12} =$

10) $5\frac{5}{6} =$

11) $7\frac{3}{5} =$

12) $3\frac{9}{10} =$

13) $6\frac{1}{3} =$

14) $5\frac{5}{6} =$

15) $8\frac{2}{5} =$

16) $4\frac{3}{8} =$

17) $3\frac{5}{9} =$

18) $2\frac{3}{14} =$

19) $7\frac{5}{6} =$

20) $5\frac{6}{7} =$

21) $4\frac{7}{8} =$

22) $3\frac{2}{9} =$

23) $2\frac{9}{11} =$

24) $10\frac{5}{7} =$

Converting improper Fractions

Convert the following improper fractions into mixed numbers

1) $\frac{55}{16} =$

2) $\frac{95}{34} =$

3) $\frac{39}{19} =$

4) $\frac{11}{3} =$

5) $\frac{61}{17} =$

6) $\frac{152}{41} =$

7) $\frac{125}{31} =$

8) $\frac{46}{7} =$

9) $\frac{43}{11} =$

10) $\frac{14}{3} =$

11) $\frac{29}{6} =$

12) $\frac{61}{15} =$

13) $\frac{56}{24} =$

14) $\frac{21}{9} =$

15) $\frac{115}{16} =$

16) $\frac{59}{6} =$

17) $\frac{144}{10} =$

18) $\frac{51}{13} =$

19) $\frac{36}{8} =$

20) $\frac{58}{6} =$

21) $\frac{9}{7} =$

22) $\frac{89}{11} =$

23) $\frac{102}{9} =$

24) $\frac{180}{19} =$

Adding Mix Numbers

Add the following fractions.

1) $3\frac{4}{9} + 2\frac{2}{9} =$

2) $3\frac{2}{5} + 2\frac{1}{5} =$

3) $1\frac{1}{7} + 2\frac{3}{7} =$

4) $4\frac{5}{6} + 2\frac{1}{3} =$

5) $1\frac{4}{15} + 2\frac{2}{5} =$

6) $4\frac{1}{3} + 1\frac{3}{4} =$

7) $3\frac{7}{9} + 3\frac{1}{6} =$

8) $3\frac{5}{8} + 2\frac{1}{2} =$

9) $3\frac{3}{4} + 2\frac{1}{4} =$

10) $1\frac{4}{11} + 2\frac{3}{11} =$

11) $4\frac{1}{2} + 2\frac{2}{5} =$

12) $5\frac{1}{4} + 2\frac{5}{6} =$

13) $6\frac{2}{7} + 2\frac{5}{7} =$

14) $3\frac{7}{8} + 2\frac{3}{16} =$

15) $3\frac{3}{4} + 3\frac{2}{9} =$

16) $4\frac{3}{5} + 2\frac{1}{6} =$

17) $7\frac{1}{2} + 6\frac{3}{5} =$

18) $6\frac{2}{3} + 1\frac{5}{12} =$

19) $2\frac{1}{9} + 7\frac{2}{3} =$

20) $4\frac{1}{6} + 2\frac{5}{9} =$

21) $5\frac{2}{3} + 6\frac{3}{4} =$

22) $7\frac{1}{8} + 1\frac{7}{24} =$

23) $5\frac{3}{7} + 4\frac{1}{8} =$

24) $8\frac{1}{3} + 4\frac{3}{4} =$

Subtracting Mix Numbers

Subtract the following fractions.

1) $8\frac{1}{4} - 7\frac{1}{4} =$

2) $5\frac{5}{6} - 5\frac{1}{6} =$

3) $9\frac{7}{11} - 8\frac{3}{11} =$

4) $5\frac{1}{2} - 2\frac{1}{6} =$

5) $4\frac{1}{4} - 1\frac{1}{8} =$

6) $9\frac{1}{3} - 5\frac{3}{7} =$

7) $5\frac{7}{9} - 2\frac{2}{9} =$

8) $8\frac{15}{17} - 5\frac{11}{17} =$

9) $9\frac{11}{14} - 3\frac{5}{14} =$

10) $7\frac{9}{10} - 6\frac{7}{10} =$

11) $8\frac{3}{4} - 5\frac{1}{12} =$

12) $6\frac{7}{8} - 3\frac{1}{8} =$

13) $7\frac{12}{35} - 3\frac{3}{7} =$

14) $6\frac{1}{3} - 4\frac{1}{9} =$

15) $10\frac{6}{7} - 7\frac{3}{7} =$

16) $9\frac{2}{3} - 3\frac{1}{3} =$

17) $5\frac{4}{11} - 3\frac{2}{11} =$

18) $7\frac{3}{10} - 5\frac{1}{5} =$

19) $8\frac{5}{6} - 5\frac{1}{12} =$

20) $3\frac{3}{4} - 3\frac{7}{16} =$

21) $8\frac{8}{13} - 3\frac{1}{3} =$

22) $6\frac{5}{6} - 4\frac{7}{30} =$

23) $5\frac{6}{7} - 4\frac{4}{11} =$

24) $6\frac{10}{19} - 2\frac{9}{19} =$

WWW.MathNotion.com

Simplify Fractions

Reduce these fractions to lowest terms

1) $\frac{18}{12} =$

2) $\frac{22}{33} =$

3) $\frac{32}{40} =$

4) $\frac{27}{36} =$

5) $\frac{8}{24} =$

6) $\frac{15}{35} =$

7) $\frac{20}{35} =$

8) $\frac{56}{70} =$

9) $\frac{9}{81} =$

10) $\frac{40}{16} =$

11) $\frac{54}{72} =$

12) $\frac{40}{120} =$

13) $\frac{12}{20} =$

14) $\frac{7}{28} =$

15) $\frac{14}{49} =$

16) $\frac{58}{87} =$

17) $\frac{72}{27} =$

18) $\frac{48}{180} =$

19) $\frac{24}{64} =$

20) $\frac{48}{42} =$

21) $\frac{120}{240} =$

22) $\frac{54}{279} =$

23) $\frac{340}{68} =$

24) $\frac{150}{600} =$

Multiplying Fractions

Find the product.

1) $\dfrac{2}{3} \times \dfrac{5}{7} =$

2) $\dfrac{3}{11} \times \dfrac{4}{9} =$

3) $\dfrac{7}{24} \times \dfrac{3}{14} =$

4) $\dfrac{7}{16} \times \dfrac{24}{35} =$

5) $\dfrac{15}{21} \times \dfrac{3}{5} =$

6) $\dfrac{18}{20} \times \dfrac{4}{9} =$

7) $\dfrac{6}{7} \times \dfrac{7}{9} =$

8) $\dfrac{54}{79} \times 0 =$

9) $\dfrac{2}{6} \times \dfrac{12}{14} =$

10) $\dfrac{24}{14} \times \dfrac{7}{8} =$

11) $\dfrac{38}{36} \times \dfrac{18}{19} =$

12) $\dfrac{8}{10} \times \dfrac{5}{64} =$

13) $\dfrac{15}{4} \times \dfrac{16}{9} =$

14) $\dfrac{25}{8} \times \dfrac{4}{10} =$

15) $\dfrac{14}{63} \times \dfrac{9}{7} =$

16) $\dfrac{12}{20} \times 4 =$

17) $\dfrac{7}{33} \times \dfrac{66}{21} =$

18) $\dfrac{5}{16} \times \dfrac{8}{10} =$

19) $\dfrac{9}{10} \times \dfrac{4}{27} =$

20) $\dfrac{6}{42} \times \dfrac{7}{12} =$

21) $\dfrac{8}{19} \times \dfrac{1}{16} =$

22) $\dfrac{10}{7} \times \dfrac{4}{80} =$

23) $\dfrac{9}{12} \times \dfrac{4}{54} =$

24) $\dfrac{60}{400} \times \dfrac{200}{600} =$

Multiplying Mixed Number

Multiply. Reduce to lowest terms.

1) $3\frac{1}{4} \times 3\frac{1}{5} =$

2) $2\frac{2}{7} \times 1\frac{1}{8} =$

3) $1\frac{1}{4} \times 2\frac{3}{5} =$

4) $2\frac{2}{9} \times 1\frac{1}{10} =$

5) $3\frac{3}{5} \times 2\frac{1}{5} =$

6) $2\frac{3}{4} \times 2\frac{2}{3} =$

7) $4\frac{1}{2} \times 1\frac{1}{9} =$

8) $2\frac{4}{5} \times 4\frac{1}{7} =$

9) $3\frac{1}{4} \times 2\frac{1}{3} =$

10) $1\frac{1}{5} \times 5\frac{1}{6} =$

11) $5\frac{1}{3} \times 2\frac{1}{8} =$

12) $2\frac{1}{3} \times 1\frac{2}{9} =$

13) $3\frac{1}{4} \times 2\frac{2}{5} =$

14) $4\frac{1}{9} \times 3\frac{1}{3} =$

15) $3\frac{1}{5} \times 2\frac{1}{7} =$

16) $4\frac{1}{2} \times 2\frac{2}{5} =$

17) $1\frac{1}{4} \times 2\frac{4}{5} =$

18) $3\frac{1}{3} \times 1\frac{1}{4} =$

19) $4\frac{4}{5} \times 1\frac{5}{7} =$

20) $6\frac{3}{4} \times 1\frac{1}{3} =$

21) $5\frac{1}{2} \times 3\frac{1}{5} =$

22) $4\frac{2}{3} \times 6\frac{1}{2} =$

Dividing Fractions

Divide these fractions.

1) $7 \div \frac{1}{2} =$

2) $\frac{9}{17} \div 9 =$

3) $\frac{3}{16} \div \frac{1}{3} =$

4) $\frac{9}{50} \div \frac{3}{10} =$

5) $\frac{3}{19} \div \frac{6}{19} =$

6) $\frac{2}{8} \div \frac{14}{32} =$

7) $0 \div \frac{1}{11} =$

8) $\frac{11}{48} \div \frac{22}{12} =$

9) $\frac{7}{24} \div \frac{21}{16} =$

10) $\frac{10}{38} \div \frac{5}{2} =$

11) $\frac{12}{25} \div \frac{18}{35} =$

12) $\frac{36}{14} \div \frac{18}{7} =$

13) $\frac{11}{27} \div \frac{11}{9} =$

14) $\frac{5}{21} \div \frac{30}{7} =$

15) $\frac{72}{37} \div \frac{8}{37} =$

16) $\frac{9}{40} \div \frac{54}{8} =$

17) $\frac{46}{9} \div \frac{23}{45} =$

18) $7 \div \frac{1}{3} =$

19) $\frac{63}{42} \div \frac{7}{6} =$

20) $\frac{4}{49} \div \frac{8}{7} =$

21) $\frac{9}{14} \div \frac{18}{21} =$

22) $\frac{7}{30} \div \frac{2}{10} =$

Dividing Mixed Number

Divide the following mixed numbers. Cancel and simplify when possible.

1) $3\frac{1}{2} \div 3\frac{1}{4} =$

2) $4\frac{1}{5} \div 2\frac{1}{3} =$

3) $6\frac{3}{5} \div 8\frac{1}{4} =$

4) $5\frac{1}{2} \div 5\frac{1}{3} =$

5) $7\frac{1}{5} \div 3\frac{1}{5} =$

6) $4\frac{3}{4} \div 1\frac{9}{10} =$

7) $6\frac{1}{4} \div 2\frac{1}{2} =$

8) $3\frac{3}{5} \div 2\frac{4}{7} =$

9) $2\frac{3}{5} \div 2\frac{2}{5} =$

10) $8\frac{4}{7} \div 1\frac{1}{9} =$

11) $5\frac{1}{3} \div 6\frac{2}{5} =$

12) $2\frac{5}{9} \div 2\frac{2}{3} =$

13) $1\frac{3}{4} \div 2\frac{1}{10} =$

14) $5\frac{2}{5} \div 4\frac{1}{2} =$

15) $3\frac{4}{6} \div 2\frac{2}{7} =$

16) $3\frac{2}{3} \div 3\frac{1}{7} =$

17) $6\frac{4}{5} \div 4\frac{1}{4} =$

18) $7\frac{1}{6} \div 3\frac{1}{2} =$

19) $5\frac{5}{8} \div 3\frac{3}{4} =$

20) $7\frac{1}{7} \div 2\frac{4}{5} =$

21) $8\frac{2}{3} \div 6\frac{1}{2} =$

22) $4\frac{2}{7} \div 5\frac{5}{8} =$

23) $9\frac{3}{4} \div 2\frac{3}{5} =$

24) $7\frac{4}{5} \div 7\frac{1}{4} =$

Comparing Fractions

Compare the fractions, and write >, < or =

1) $\dfrac{13}{2}$ _____ $\dfrac{19}{10}$

2) $\dfrac{15}{4}$ _____ $\dfrac{5}{7}$

3) $\dfrac{5}{8}$ _____ $\dfrac{4}{5}$

4) $\dfrac{11}{3}$ _____ $\dfrac{12}{8}$

5) $\dfrac{2}{9}$ _____ $\dfrac{4}{7}$

6) $\dfrac{13}{5}$ _____ $\dfrac{17}{4}$

7) $\dfrac{14}{9}$ _____ $\dfrac{8}{11}$

8) $\dfrac{15}{13}$ _____ $\dfrac{21}{8}$

9) $5\dfrac{1}{10}$ _____ $8\dfrac{1}{15}$

10) $9\dfrac{1}{12}$ _____ $7\dfrac{1}{9}$

11) $4\dfrac{1}{4}$ _____ $4\dfrac{1}{7}$

12) $8\dfrac{6}{7}$ _____ $8\dfrac{3}{8}$

13) $1\dfrac{5}{9}$ _____ $4\dfrac{2}{3}$

14) $\dfrac{1}{24}$ _____ $\dfrac{2}{13}$

15) $\dfrac{42}{23}$ _____ $\dfrac{29}{72}$

16) $\dfrac{14}{200}$ _____ $\dfrac{8}{81}$

17) $19\dfrac{1}{3}$ _____ $19\dfrac{1}{7}$

18) $\dfrac{1}{8}$ _____ $\dfrac{1}{12}$

19) $\dfrac{1}{11}$ _____ $\dfrac{1}{15}$

20) $\dfrac{1}{15}$ _____ $\dfrac{7}{12}$

21) $\dfrac{10}{33}$ _____ $\dfrac{8}{59}$

22) $\dfrac{6}{7}$ _____ $\dfrac{3}{8}$

23) $6\dfrac{2}{5}$ _____ $4\dfrac{12}{5}$

24) $2\dfrac{12}{5}$ _____ $3\dfrac{4}{5}$

Answer key Chapter 4

Adding Fractions – Like Denominator

1) $\frac{2}{3}$
2) $\frac{4}{7}$
3) $\frac{7}{9}$
4) $\frac{8}{15}$
5) $\frac{9}{23}$
6) $\frac{15}{29}$
7) $\frac{8}{19}$
8) $\frac{3}{8}$
9) $\frac{13}{31}$
10) $\frac{13}{51}$
11) $\frac{4}{17}$
12) $\frac{6}{13}$
13) $\frac{24}{41}$
14) $\frac{1}{5}$
15) $\frac{13}{21}$
16) $\frac{4}{11}$
17) $\frac{6}{11}$
18) $\frac{25}{67}$
19) $\frac{9}{19}$
20) $\frac{33}{47}$

Adding Fractions – Unlike Denominator

1) $\frac{2}{3}$
2) $\frac{17}{12}$
3) $\frac{11}{10}$
4) $\frac{31}{30}$
5) $\frac{7}{15}$
6) $\frac{1}{2}$
7) $\frac{13}{21}$
8) $\frac{1}{4}$
9) $\frac{3}{4}$
10) $\frac{7}{8}$
11) $\frac{11}{21}$
12) $\frac{13}{36}$
13) $\frac{19}{35}$
14) $\frac{5}{33}$
15) $\frac{19}{27}$
16) $\frac{3}{5}$
17) $\frac{19}{45}$
18) $\frac{19}{35}$
19) $\frac{8}{11}$
20) $\frac{7}{16}$
21) $\frac{17}{48}$
22) $\frac{13}{30}$
23) $\frac{13}{42}$
24) $\frac{17}{72}$

Subtracting Fractions – Like Denominator

1) 1
2) $\frac{1}{4}$
3) $\frac{3}{13}$
4) 2
5) $\frac{2}{21}$
6) $\frac{8}{43}$
7) $\frac{6}{23}$
8) $\frac{3}{47}$
9) $\frac{1}{5}$
10) $\frac{17}{48}$
11) $\frac{4}{7}$
12) $\frac{8}{51}$
13) $\frac{4}{11}$
14) $\frac{15}{49}$
15) $\frac{6}{17}$
16) $\frac{6}{23}$
17) $\frac{4}{7}$
18) $\frac{4}{31}$

SBAC Math Practice Grade 5

19) $\frac{9}{61}$ 21) $\frac{5}{63}$ 23) $\frac{24}{77}$

20) $\frac{11}{52}$ 22) $\frac{30}{83}$ 24) $\frac{12}{55}$

Subtracting Fractions – Unlike Denominator

1) $\frac{1}{6}$ 8) $\frac{1}{30}$ 15) $\frac{3}{8}$

2) $\frac{22}{45}$ 9) $\frac{11}{35}$ 16) $\frac{13}{63}$

3) $\frac{1}{12}$ 10) $\frac{52}{99}$ 17) $\frac{5}{48}$

4) $\frac{43}{90}$ 11) $\frac{29}{48}$ 18) $\frac{13}{40}$

5) $\frac{17}{24}$ 12) $\frac{1}{20}$ 19) $\frac{19}{56}$

6) $\frac{33}{80}$ 13) $\frac{28}{45}$ 20) $\frac{3}{11}$

7) $\frac{7}{15}$ 14) $\frac{39}{70}$

Converting Mix Numbers

1) $\frac{13}{5}$ 9) $\frac{25}{12}$ 17) $\frac{32}{9}$

2) $\frac{19}{5}$ 10) $\frac{35}{6}$ 18) $\frac{31}{14}$

3) $\frac{23}{4}$ 11) $\frac{38}{5}$ 19) $\frac{47}{6}$

4) $\frac{29}{8}$ 12) $\frac{39}{10}$ 20) $\frac{41}{7}$

5) $\frac{32}{5}$ 13) $\frac{19}{3}$ 21) $\frac{39}{8}$

6) $\frac{71}{9}$ 14) $\frac{35}{6}$ 22) $\frac{29}{9}$

7) $\frac{32}{15}$ 15) $\frac{42}{5}$ 23) $\frac{31}{11}$

8) $\frac{40}{13}$ 16) $\frac{35}{8}$ 24) $\frac{75}{7}$

Converting improper Fractions

1) $3\frac{7}{16}$ 6) $3\frac{29}{41}$ 11) $4\frac{5}{6}$

2) $2\frac{27}{34}$ 7) $4\frac{1}{31}$ 12) $4\frac{1}{15}$

3) $2\frac{1}{19}$ 8) $6\frac{4}{7}$ 13) $2\frac{1}{3}$

4) $3\frac{2}{3}$ 9) $3\frac{10}{11}$ 14) $2\frac{1}{3}$

5) $3\frac{10}{17}$ 10) $4\frac{2}{3}$ 15) $7\frac{3}{16}$

16) $9\frac{5}{6}$

17) $14\frac{2}{5}$

18) $3\frac{12}{13}$

19) $4\frac{1}{2}$

20) $9\frac{2}{3}$

21) $1\frac{2}{7}$

22) $8\frac{1}{11}$

23) $11\frac{1}{3}$

24) $9\frac{9}{19}$

Adding Mix Numbers

1) $5\frac{2}{3}$

2) $5\frac{3}{5}$

3) $3\frac{4}{7}$

4) $7\frac{1}{6}$

5) $3\frac{2}{3}$

6) $6\frac{1}{12}$

7) $6\frac{17}{18}$

8) $6\frac{1}{8}$

9) 6

10) $3\frac{7}{11}$

11) $6\frac{9}{10}$

12) $8\frac{1}{12}$

13) 9

14) $6\frac{1}{16}$

15) $6\frac{35}{36}$

16) $6\frac{23}{30}$

17) $14\frac{1}{10}$

18) $8\frac{1}{12}$

19) $9\frac{7}{9}$

20) $6\frac{13}{18}$

21) $12\frac{5}{12}$

22) $8\frac{5}{12}$

23) $9\frac{31}{56}$

24) $13\frac{1}{12}$

Subtracting Mix Numbers

1) 1

2) $\frac{2}{3}$

3) $1\frac{4}{11}$

4) $3\frac{1}{3}$

5) $3\frac{1}{8}$

6) $3\frac{19}{21}$

7) $3\frac{5}{9}$

8) $3\frac{4}{17}$

9) $6\frac{3}{7}$

10) $1\frac{1}{5}$

11) $3\frac{2}{3}$

12) $3\frac{3}{4}$

13) $3\frac{32}{35}$

14) $2\frac{2}{9}$

15) $3\frac{3}{7}$

16) $6\frac{1}{3}$

17) $2\frac{2}{11}$

18) $2\frac{1}{10}$

19) $3\frac{3}{4}$

20) $\frac{5}{16}$

21) $5\frac{11}{39}$

22) $2\frac{3}{5}$

23) $1\frac{38}{77}$

24) $4\frac{1}{19}$

Simplify Fractions

1) $\frac{3}{2}$

2) $\frac{2}{3}$

3) $\frac{4}{5}$

SBAC Math Practice Grade 5

4) $\frac{3}{4}$

5) $\frac{1}{3}$

6) $\frac{3}{7}$

7) $\frac{4}{7}$

8) $\frac{4}{5}$

9) $\frac{1}{9}$

10) $\frac{5}{2}$

11) $\frac{3}{4}$

12) $\frac{1}{3}$

13) $\frac{3}{5}$

14) $\frac{1}{4}$

15) $\frac{2}{7}$

16) $\frac{2}{3}$

17) $\frac{8}{3}$

18) $\frac{4}{15}$

19) $\frac{3}{8}$

20) $\frac{8}{7}$

21) $\frac{1}{2}$

22) $\frac{6}{31}$

23) 5

24) $\frac{1}{4}$

Multiplying Fractions

1) $\frac{10}{21}$

2) $\frac{4}{33}$

3) $\frac{1}{16}$

4) $\frac{3}{10}$

5) $\frac{3}{7}$

6) $\frac{2}{5}$

7) $\frac{2}{3}$

8) 0

9) $\frac{2}{7}$

10) $\frac{3}{2}$

11) 1

12) $\frac{1}{16}$

13) $\frac{20}{3}$

14) $\frac{5}{4}$

15) $\frac{2}{7}$

16) $\frac{12}{5}$

17) $\frac{2}{3}$

18) $\frac{1}{4}$

19) $\frac{2}{15}$

20) $\frac{1}{12}$

21) $\frac{1}{38}$

22) $\frac{1}{14}$

23) $\frac{1}{18}$

24) $\frac{1}{20}$

Multiplying Mixed Number

1) $10\frac{2}{5}$

2) $2\frac{4}{7}$

3) $3\frac{1}{4}$

4) $2\frac{4}{9}$

5) $7\frac{23}{25}$

6) $7\frac{1}{3}$

7) 5

8) $11\frac{3}{5}$

9) $7\frac{7}{12}$

10) $6\frac{1}{5}$

11) $11\frac{1}{3}$

12) $2\frac{23}{27}$

13) $7\frac{4}{5}$

14) $13\frac{19}{27}$

15) $6\frac{6}{7}$

16) $10\frac{4}{5}$

17) $3\frac{1}{2}$

18) $4\frac{1}{6}$

19) $8\frac{8}{35}$

20) 9

21) $17\frac{3}{5}$

22) $30\frac{1}{3}$

Dividing Fractions

1) 14
2) $\frac{1}{17}$
3) $\frac{9}{16}$
4) $\frac{3}{5}$
5) $\frac{1}{2}$
6) $\frac{4}{7}$
7) 0
8) $\frac{1}{8}$
9) $\frac{2}{9}$
10) $\frac{2}{19}$
11) $\frac{14}{15}$
12) 1
13) $\frac{1}{3}$
14) $\frac{1}{18}$
15) 9
16) $\frac{1}{30}$
17) 10
18) 21
19) $\frac{9}{7}$
20) $\frac{1}{14}$
21) $\frac{3}{4}$
22) $\frac{7}{6}$

Dividing Mixed Number

1) $1\frac{1}{13}$
2) $1\frac{4}{5}$
3) $\frac{4}{5}$
4) $1\frac{1}{32}$
5) $2\frac{1}{4}$
6) $2\frac{1}{2}$
7) $2\frac{1}{2}$
8) $1\frac{2}{5}$
9) $1\frac{1}{12}$
10) $7\frac{5}{7}$
11) $\frac{5}{6}$
12) $\frac{23}{24}$
13) $\frac{5}{6}$
14) $1\frac{1}{5}$
15) $1\frac{29}{48}$
16) $1\frac{1}{6}$
17) $1\frac{3}{5}$
18) $2\frac{1}{21}$
19) $1\frac{1}{2}$
20) $2\frac{27}{49}$
21) $1\frac{1}{3}$
22) $\frac{16}{21}$
23) $3\frac{3}{4}$
24) $1\frac{11}{145}$

Comparing Fractions

1) >
2) >
3) <
4) >
5) <
6) <
7) >
8) <
9) <
10) >
11) >
12) >
13) <
14) <
15) >
16) <
17) >
18) >
19) >
20) <
21) >
22) >
23) =
24) >

Chapter 5: Decimal

Graph Decimals

Write the decimals indicated by the arrows.

1)

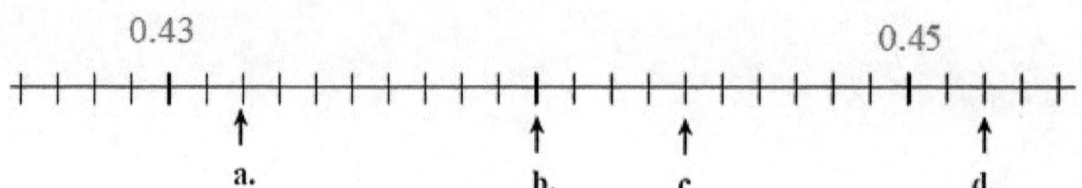

a. _____ b. _____ c. _____ d. _____

2)

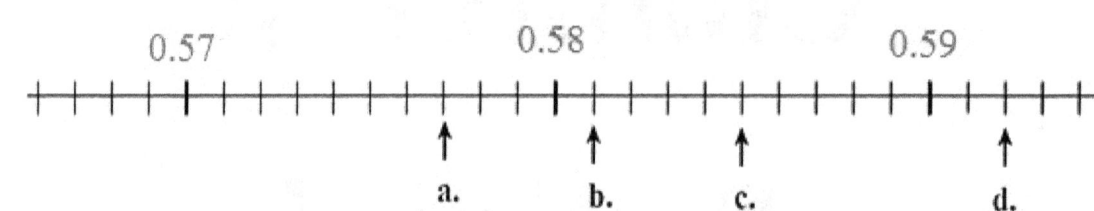

a. _____ b. _____ c. _____ d. _____

3)

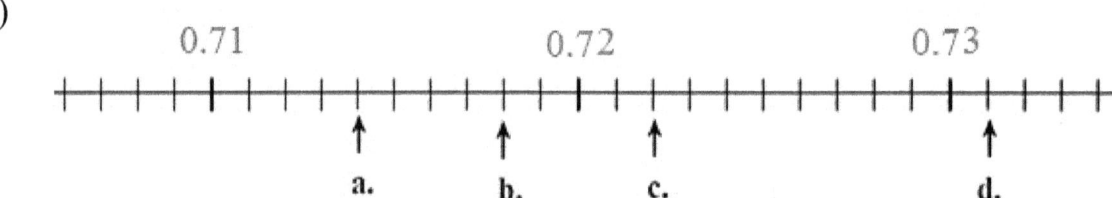

a. _____ b. _____ c. _____ d. _____

4)

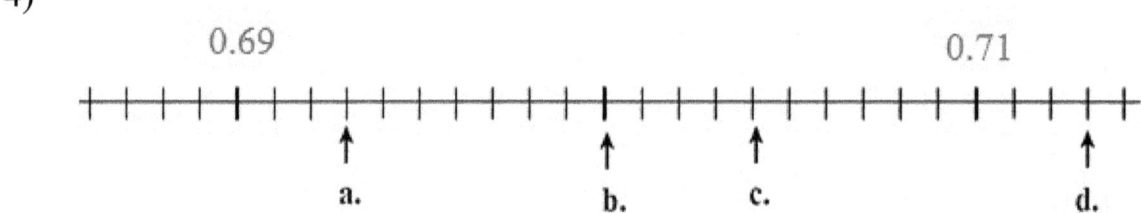

a. _____ b. _____ c. _____ d. _____

Round Decimals

Round each number to the correct place value

1) 0.9̲2 =

2) 4.2̲1 =

3) 8.8̲33 =

4) 0.5̲69 =

5) 7̲.779 =

6) 0.04̲9 =

7) 9.3̲7 =

8) 25.33̲1 =

9) 4.62̲9 =

10) 10.4̲71 =

11) 3.2̲6 =

12) 4.2̲29 =

13) 6.2̲18 =

14) 9.16̲24 =

15) 36̲.29 =

16) 47̲.68 =

17) 63̲2.785 =

18) 577.8̲29 =

19) 24.57̲9 =

20) 83̲.91 =

21) 5.41̲35 =

22) 86.2̲76 =

23) 354.3̲39 =

24) 0.92̲66 =

25) 0.007̲4 =

26) 8.044̲9 =

27) 41.65̲96 =

28) 19.08̲27 =

Decimals Addition

Add the following.

1) 32.19 + 31.16

2) 0.54 + 0.31

3) 21.42 + 12.57

4) 57.189 + 7.221

5) 22.740 + 8.37

6) 6.712 + 4.105

7) 78.46 + 18.54

8) 66.24 + 20.36

9) 39.11 + 15.09

10) 4.88 + 19.45

11) 16.254 + 34.227

12) 36.66 + 4.82

13) 42.45 + 9.37

14) 124.41 + 5.71

Decimals Subtraction

Subtract the following

1) 8.19
   -2.46

2) 46.25
   -16.18

3) 0.89
   -0.5

4) 24.354
   -7.8

5) 0.789
   -0.06

6) 63.25
   -42.52

7) 157.75
   -94.87

8) 45.65
   -19.57

9) 65.42
   -47.71

10) 8.652
    -0.553

11) 48.61
    -29.52

12) 16.399
    -5.389

13) 35.251
    -9.169

14) 148.42
    -11.78

Decimals Multiplication

Solve.

1) 2.4 × 1.9

2) 3.6 × 5.2

3) 4.02 × 2.04

4) 45.9 × 10

5) 49.8 × 100

6) 41.56 × 4.2

7) 24.51 × 10.2

8) 1.89 × 7.35

9) 14.05 × 0.09

10) 51.03 × 4.04

11) 14.56 × 12.4

12) 9.56 × 0.04

13) 7.5 × 0.11

14) 23.1 × 4.02

Decimal Division

Dividing Decimals.

1) $7 \div 1{,}000 =$

2) $3 \div 10{,}000 =$

3) $2.8 \div 10 =$

4) $0.08 \div 100 =$

5) $5 \div 25 =$

6) $4 \div 48 =$

7) $8 \div 40 =$

8) $7 \div 140 =$

9) $9 \div 10{,}000 =$

10) $0.6 \div 0.54 =$

11) $0.4 \div 0.004 =$

12) $0.3 \div 0.15 =$

13) $0.7 \div 0.56 =$

14) $0.6 \div 0.0006 =$

15) $4.9 \div 100 =$

16) $7.3 \div 100 =$

17) $8.5 \div 10 =$

18) $16.2 \div 4.4 =$

19) $36.9 \div 3.3 =$

20) $0.8 \div 0.08 =$

21) $8.04 \div 4.2 =$

22) $0.09 \div 0.30 =$

23) $0.8 \div 6.4 =$

24) $0.07 \div 42 =$

25) $6.28 \div 0.6 =$

26) $0.026 \div 13 =$

Comparing Decimals

Write the Correct Comparison Symbol (>, < or =)

1) 1.98 ____ 3.12

2) 0.6 ____ 0.549

3) 19.01 ____ 19.010

4) 5.05 ____ 5.50

5) 0.811 ____ 0.81

6) 0.658 ____ 0.865

7) 5.46 ____ 5.391

8) 7.021 ____ 7.035

9) 56.321 ____ 56.123

10) 4.69 ____ 4.069

11) 3.55 ____ 3.555

12) 0.08 ____ 0.12

13) 2.405 ____ 2.45

14) 7.53 ____ 7.35

15) 0.11 ____ 0.011

16) 86.09 ____ 86.090

17) 0.190 ____ 0.21

18) 53.98 ____ 54.07

19) 0.072 ____ 0.720

20) 43.2 ____ 34.9

21) 17.99 ____ 20.19

22) 0.087 ____ 0.0807

23) 3.059 ____ 0.3059

24) 8.2 ____ 0.825

25) 8.77 ____ 0.877

26) 9.56 ____ 9.5600

27) 4.97 ____ 0.497

28) 3.0504 ____ 3.0540

Convert Fraction to Decimal

Write each as a decimal.

1) $\frac{6}{10} =$

2) $\frac{58}{100} =$

3) $\frac{80}{100} =$

4) $\frac{5}{40} =$

5) $\frac{4}{50} =$

6) $\frac{7}{100} =$

7) $\frac{5}{80} =$

8) $\frac{21}{84} =$

9) $\frac{36}{400} =$

10) $\frac{3}{11} =$

11) $\frac{24}{48} =$

12) $\frac{18}{48} =$

13) $\frac{6}{20} =$

14) $\frac{9}{125} =$

15) $\frac{36}{120} =$

16) $\frac{15}{20} =$

17) $\frac{73}{100} =$

18) $\frac{9}{45} =$

19) $\frac{9}{10} =$

20) $\frac{6}{60} =$

21) $\frac{7}{42} =$

22) $\frac{11}{88} =$

Answer key Chapter 5

Graph Decimals

1) a. 0.432   b. 0.44   c. 0.444   d. 0.452
2) a. 0.577   b. 0.581   c. 0.585   d. 0592
3) a. 0.714   b. 0.718   c. 0.722   d. 0.731
4) a. 0.693   b. 0.70   c. 0.704   d. 0.713

Round Decimals

1) 0.9
2) 4.2
3) 8.8
4) 0.6
5) 8.0
6) 0.05
7) 9.4
8) 25.33
9) 4.63
10) 10.5
11) 3.3
12) 4.2
13) 6.2
14) 9.16
15) 36.0
16) 48.0
17) 630.0
18) 577.8
19) 24.58
20) 84.0
21) 5.41
22) 86.3
23) 354.3
24) 0.93
25) 0.007
26) 8.04
27) 41.66
28) 19.08

Decimals Addition

1) 63.35
2) 0.85
3) 33.99
4) 64.41
5) 31.11
6) 10.817
7) 97
8) 86.6
9) 54.2
10) 24.33
11) 50.481
12) 41.48
13) 51.82
14) 130.12

Decimals Subtraction

1) 5.73
2) 30.07
3) 0.39
4) 16.554
5) 0.729
6) 20.73
7) 62.88
8) 26.08
9) 17.71
10) 8.099
11) 19.09
12) 11.01
13) 26.082
14) 136.64

Decimals Multiplication

1) 4.56
2) 18.72
3) 8.2008

SBAC Math Practice Grade 5

4) 459
5) 4,980
6) 174.552
7) 250.002
8) 13.8915
9) 1.2645
10) 206.1612
11) 180.544
12) 0.3824
13) 0.825
14) 92.862

Decimal Division

1) 0.007
2) 0.0003
3) 0.28
4) 0.0008
5) 0.2
6) 0.833….
7) 0.2
8) 0.05
9) 0.0009
10) 1.111…
11) 100
12) 2
13) 1.25
14) 1,000
15) 0.049
16) 0.073
17) 0.85
18) 3.68181…
19) 11.1818…
20) 10
21) 1.19428…
22) 0.3
23) 0.125
24) 0.00167
25) 10.467
26) 0.002

Comparing Decimals

1) <
2) >
3) =
4) <
5) >
6) <
7) >
8) <
9) >
10) >
11) <
12) <
13) <
14) >
15) >
16) =
17) <
18) <
19) <
20) >
21) <
22) >
23) >
24) >
25) >
26) =
27) >
28) <

Convert Fraction to Decimal

1) 0.6
2) 0.58
3) 0.8
4) 0.125
5) 0.08
6) 0.07
7) 0.0625
8) 0.25
9) 0.09
10) 0.27
11) 0.5
12) 0.375
13) 0.3
14) 0.072
15) 0.3

WWW.MathNotion.com

16) 0.75   19) 0.9   22) 0.125
17) 0.73   20) 0.1
18) 0.2   21) 0.167

Chapter 6:

Exponent and

Radicals

Positive Exponents

Simplify. Your answer should contain only positive exponents.

1) $4^3 =$

2) $5^3 =$

3) $\frac{7yx}{y} =$

4) $(10x^3 x)^2 =$

5) $(x^4)^3 =$

6) $(\frac{1}{2})^5 =$

7) $0^5 =$

8) $7 \times 7 \times 7 =$

9) $2 \times 2 \times 2 \times 2 \times 2 =$

10) $(3x^3 y)^3 =$

11) $3^4 =$

12) $(2x^2 y)^4 =$

13) $6 \times 10^3 =$

14) $0.5 \times 0.5 \times 0.5 =$

15) $\frac{1}{9} \times \frac{1}{9} \times \frac{1}{9} =$

16) $6^3 =$

17) $(4x^6 y^3)^2 =$

18) $2^6 =$

19) $z \times z \times z =$

20) $9 \times 9 \times 9 \times 9 \times 9 =$

21) $(3x^3 y^3 z)^2 =$

22) $5^0 =$

23) $(12x^3 y^{-2})^2 =$

24) $(3x^3 y^2)^4 =$

Negative Exponents

Simplify. Leave no negative exponents.

1) $3^{-3} =$

2) $8^{-2} =$

3) $(\frac{1}{4})^{-3} =$

4) $7^{-2} =$

5) $1^{120} =$

6) $6^{-4} =$

7) $(\frac{1}{3})^{-4} =$

8) $-12y^{-9} =$

9) $(\frac{1}{y^{-4}})^{-6} =$

10) $x^{-\frac{6}{7}} =$

11) $\frac{1}{9^{-4}} =$

12) $2^{-6} =$

13) $3^{-5} =$

14) $12^{-2} =$

15) $20^{-2} =$

16) $x^{-9} =$

17) $(x^3)^{-5} =$

18) $x^{-3} \times x^{-3} \times x^{-3} =$

19) $\frac{1}{8} \times \frac{1}{8} \times \frac{1}{8} =$

20) $10^{-4} =$

21) $200z^{-5} =$

22) $2^{-8} =$

23) $(-\frac{1}{13})^2 =$

24) $23^0 =$

25) $(\frac{1}{x})^{-31} =$

26) $16^{-2} =$

Add and subtract Exponents

Solve each problem.

1) $3^3 + 4^2 =$

2) $x^9 + 2x^9 =$

3) $7b^4 - 6b^4 =$

4) $8 + 6^2 =$

5) $12 - 5^2 =$

6) $18 + 2^3 =$

7) $6x^2 + 9x^2 =$

8) $7^2 + 3^4 =$

9) $4^4 - 3^3 =$

10) $9^2 - 12^0 =$

11) $8^2 - 5^2 =$

12) $11^2 + 2^4 =$

13) $13^2 - 9^2 =$

14) $10^2 + 11^2 =$

15) $5^3 - 2^4 =$

16) $1^{32} + 1^{12} =$

17) $8^2 - 7^2 =$

18) $3^5 - 3^4 =$

19) $10^2 - 2^3 =$

20) $6^2 + 3^3 =$

21) $5^3 + 2^4 =$

22) $24 + 3^2 =$

23) $15x^7 + 4x^7 =$

24) $7^0 + 9^2 =$

25) $6^2 + 6^2 =$

26) $4^2 + 2^5 =$

27) $(\frac{1}{5})^2 + (\frac{1}{5})^2 =$

28) $7^2 + 6^2 =$

Exponent multiplication

Simplify each of the following

1) $4^4 \times 4^5 =$

2) $9^2 \times 10^0 =$

3) $12^2 \times 2^2 =$

4) $a^{-7} \times a^{-7} =$

5) $y^{-6} \times y^{-6} \times y^{-6} =$

6) $4^6 \times 5^9 \times 4^{-5} \times 5^{-8} =$

7) $5x^2y^4 \times 3x^2y^5 =$

8) $(x^4)^2 =$

9) $(x^3y^4)^4 \times (x^3y^2)^{-4} =$

10) $9^5 \times 9^3 =$

11) $z^{6b} \times z^0 =$

12) $8^5 \times 8^4 =$

13) $a^{7m} \times a^{5n} =$

14) $3a^n \times 5b^n =$

15) $7^{-4} \times 6^{-4} =$

16) $9^8 \times 5^8 =$

17) $(5^7)^4 =$

18) $(\frac{1}{8})^3 \times (\frac{1}{8})^6 \times (\frac{1}{8})^4 =$

19) $(\frac{1}{6})^{41} \times 6^{41} =$

20) $(5m)^{\frac{2}{7}} \times (-4m)^{\frac{2}{7}} =$

21) $(x^3y)^{\frac{1}{3}} \times (xy^4)^{\frac{1}{3}} =$

22) $(3a^{2m}b^n)^r =$

23) $(6x^4y^4)^2 =$

24) $(x^{\frac{1}{5}}y^3)^{\frac{-1}{5}} \times (x^7y^8)^0 =$

25) $9^4 \times 9^7 =$

26) $16^{\frac{3}{8}} \times 16^{\frac{1}{8}} =$

27) $8^6 \times 2^6 =$

28) $(x^{18})^0 =$

Exponent division

Simplify. Your answer should contain only positive exponents.

1) $\dfrac{7^5}{7} =$

2) $\dfrac{21x^3}{x} =$

3) $\dfrac{a^{3m}}{a^{4n}} =$

4) $\dfrac{2x^{-7}}{14x^{-2}} =$

5) $\dfrac{54x^7}{9x^3} =$

6) $\dfrac{11x^8}{3x^9} =$

7) $\dfrac{45x^9}{15y^5} =$

8) $\dfrac{25xy^9}{x^5y^3} =$

9) $\dfrac{2x^7}{5x} =$

10) $\dfrac{36x^6y^7}{4x^7} =$

11) $\dfrac{9x^4}{27x^9y^{10}} =$

12) $\dfrac{5yx^5}{35yx^{12}} =$

13) $\dfrac{32x^4y^4}{8x^3y^5} =$

14) $\dfrac{x^{2.25}}{x^{0.25}} =$

15) $\dfrac{7x^5y}{28xy^4} =$

16) $\dfrac{36b^4r^5}{18a^3b^6} =$

17) $\dfrac{40x^5}{20x^7} =$

18) $\dfrac{24x^2}{6x^5} =$

19) $\dfrac{7^9}{7^7} =$

20) $\dfrac{2x}{x^{13}} =$

21) $\dfrac{15^9}{15^6} =$

22) $\dfrac{4xy^6}{16y^4} =$

23) $\dfrac{12x^4y}{144xy^5} =$

24) $\dfrac{42x^7}{7y^8} =$

Scientific Notation

Write each number in scientific notation.

1) 8,700,000 =

2) 6,000 =

3) 0.0005 =

4) 250,000 =

5) 0.0159 =

6) 0.056 =

7) 0.000005 =

8) 12,000,000 =

9) 800,000 =

10) 12,100,000 =

11) 0.0009 =

12) 0.000008 =

13) 0.00055 =

14) 600,000 =

15) 3,054 =

16) 1,570,000 =

17) 0.00013 =

18) 0.4 =

19) 0.97 =

20) 400,000 =

21) 4,150,000 =

22) 0.0048 =

23) 0.00027 =

24) 3,125 =

25) 63,887 =

26) 121,100 =

27) 54,200 =

28) 6,250,000 =

Square Roots

Find the square root of each number.

1) $\sqrt{49} =$

2) $\sqrt{1} =$

3) $\sqrt{25} =$

4) $\sqrt{121} =$

5) $\sqrt{400} =$

6) $\sqrt{36} =$

7) $\sqrt{0} =$

8) $\sqrt{100} =$

9) $\sqrt{16} =$

10) $\sqrt{81} =$

11) $\sqrt{144} =$

12) $\sqrt{900} =$

13) $\sqrt{64} =$

14) $\sqrt{225} =$

15) $\sqrt{9} =$

16) $\sqrt{169} =$

17) $\sqrt{1,600} =$

18) $\sqrt{0.09} =$

19) $\sqrt{625} =$

20) $\sqrt{256} =$

21) $\sqrt{1.44} =$

22) $\sqrt{4,900} =$

23) $\sqrt{484} =$

24) $\sqrt{6,400} =$

25) $\sqrt{289} =$

26) $\sqrt{2.56} =$

27) $\sqrt{361} =$

28) $\sqrt{0.01} =$

Simplify Square Roots

Simplify the following.

1) $\sqrt{80} =$

2) $\sqrt{72} =$

3) $\sqrt{20} =$

4) $\sqrt{90} =$

5) $\sqrt{288} =$

6) $\sqrt{125} =$

7) $4\sqrt{20} =$

8) $5\sqrt{200} =$

9) $\sqrt{32} =$

10) $4\sqrt{8} =$

11) $5\sqrt{7} + 6\sqrt{7} =$

12) $\frac{16}{5+\sqrt{3}} =$

13) $\sqrt{98} =$

14) $\frac{7}{4-\sqrt{3}} =$

15) $\sqrt{12} \times \sqrt{3} =$

16) $\frac{\sqrt{500}}{\sqrt{5}} =$

17) $\frac{\sqrt{60}}{\sqrt{15 \times 4}} =$

18) $\sqrt{180y^8} =$

19) $5\sqrt{49a} =$

20) $\sqrt{58 + 6} + \sqrt{16} =$

21) $\sqrt{162} =$

22) $\sqrt{242} =$

23) $\sqrt{175} =$

24) $\sqrt{147} =$

25) $\sqrt{512} =$

26) $\sqrt{112} =$

Answer key Chapter 6

Positive Exponents

1) 64
2) 125
3) $7x$
4) $100x^8$
5) x^{12}
6) $\frac{1}{32}$
7) 0
8) 7^3
9) 2^5
10) $27x^9y^3$
11) 81
12) $16x^8y^4$
13) 6,000
14) 0.5^3
15) $(\frac{1}{9})^3$
16) 216
17) $16x^{12}y^6$
18) 64
19) z^3
20) 9^5
21) $9x^6y^6z^2$
22) 1
23) $\frac{144x^6}{y^4}$
24) $81x^{12}y^8$

Negative Exponents

1) $\frac{1}{27}$
2) $\frac{1}{64}$
3) 64
4) $\frac{1}{49}$
5) 1
6) $\frac{1}{1,296}$
7) 81
8) $\frac{-12}{y^9}$
9) $\frac{1}{y^{24}}$
10) $\frac{1}{x^{\frac{6}{7}}}$
11) 9^4
12) $\frac{1}{64}$
13) $\frac{1}{243}$
14) $\frac{1}{144}$
15) $\frac{1}{400}$
16) $\frac{1}{x^9}$
17) $\frac{1}{x^{15}}$
18) $\frac{1}{x^9}$
19) $\frac{1}{8^3}$
20) $\frac{1}{10,000}$
21) $\frac{200}{z^5}$
22) $\frac{1}{256}$
23) $\frac{1}{169}$
24) 1
25) x^{31}
26) $\frac{1}{256}$

Add and subtract Exponents

1) 43
2) $3x^9$
3) b^4
4) 44
5) -13
6) 26
7) $15x^2$
8) 130
9) 229
10) 80
11) 39
12) 137
13) 88
14) 221
15) 109

SBAC Math Practice Grade 5

16) 2
17) 15
18) 162
19) 92
20) 63
21) 141
22) 33
23) $19x^7$
24) 82
25) 72
26) 48
27) $\frac{2}{25}$
28) 85

Exponent multiplication

1) 4^9
2) 81
3) 576
4) a^{-14}
5) y^{-18}
6) 20
7) $15x^4y^9$
8) x^8
9) y^8
10) 9^8
11) Z^{6b}
12) 8^9
13) a^{7m+5n}
14) $15(ab)^n$
15) 42^{-4}
16) 45^8
17) 5^{28}
18) $(\frac{1}{8})^{13}$
19) 1
20) $(-20m^2)^{\frac{2}{7}}$
21) $x^{\frac{10}{3}}y^{\frac{5}{3}}$
22) $3^r a^{2mr} b^{nr}$
23) $36x^8y^8$
24) $x^{\frac{-1}{25}}y^{\frac{-3}{5}}$
25) 9^{11}
26) $16^{\frac{1}{2}}$
27) $16^6 = 2^{24}$
28) 1

Exponent division

1) 7^4
2) $21x^2$
3) a^{3m-4n}
4) $\frac{1}{7x^5}$
5) $6x^4$
6) $\frac{11}{3x}$
7) $\frac{3x^9}{y^5}$
8) $\frac{25y^6}{x^4}$
9) $\frac{2x^6}{5}$
10) $\frac{9y^7}{x}$
11) $\frac{1}{3x^5y^{10}}$
12) $\frac{1}{7x^7}$
13) $\frac{4x}{y}$
14) x^2
15) $\frac{x^4}{4y^3}$
16) $\frac{2r^5}{a^3b^2}$
17) $\frac{2}{x^2}$
18) $\frac{4}{x^3}$
19) 7^2
20) $\frac{2}{x^{12}}$
21) 15^3
22) $\frac{1}{4}xy^2$
23) $\frac{x^3}{12y^4}$
24) $\frac{6x^7}{y^8}$

Scientific Notation

1) 8.7×10^6
2) 6×10^3
3) 5×10^{-4}
4) 2.5×10^5
5) 1.59×10^{-2}
6) 5.6×10^{-2}

SBAC Math Practice Grade 5

7) 5×10^{-6}
8) 1.2×10^7
9) 8×10^5
10) 1.21×10^7
11) 9×10^{-4}
12) 8×10^{-6}
13) 5.5×10^{-4}
14) 6×10^5

15) 3.054×10^3
16) 1.57×10^6
17) 1.3×10^{-4}
18) 4×10^{-1}
19) 9.7×10^{-2}
20) 4×10^5
21) 4.15×10^6
22) 4.8×10^{-3}

23) 2.7×10^{-4}
24) 3.125×10^3
25) 6.3887×10^4
26) 1.211×10^5
27) 5.42×10^4
28) 6.25×10^6

Square Roots

1) 7
2) 1
3) 5
4) 11
5) 20
6) 6
7) 0
8) 10
9) 4
10) 9

11) 12
12) 30
13) 8
14) 15
15) 3
16) 13
17) 40
18) 0.3
19) 25
20) 16

21) 1.2
22) 70
23) 22
24) 80
25) 17
26) 1.6
27) 19
28) 0.1

Simplify Square Roots

1) $4\sqrt{5}$
2) $6\sqrt{2}$
3) $2\sqrt{5}$
4) $3\sqrt{10}$
5) $12\sqrt{2}$
6) $5\sqrt{5}$
7) $8\sqrt{5}$
8) $50\sqrt{2}$
9) $4\sqrt{2}$

10) $8\sqrt{2}$
11) $11\sqrt{7}$
12) $\frac{8}{11}(5 - \sqrt{3})$
13) $7\sqrt{2}$
14) $4 - \sqrt{3}$
15) 6
16) 10
17) 1
18) $6y^4\sqrt{5}$

19) $35\sqrt{a}$
20) 12
21) $9\sqrt{2}$
22) $11\sqrt{2}$
23) $5\sqrt{7}$
24) $7\sqrt{3}$
25) $16\sqrt{2}$
26) $4\sqrt{7}$

Chapter 7: Ratio, Proportion and Percent

Proportions

Find a missing number in a proportion.

1) $\dfrac{5}{3} = \dfrac{35}{a}$

2) $\dfrac{a}{6} = \dfrac{16}{24}$

3) $\dfrac{11}{44} = \dfrac{a}{4}$

4) $\dfrac{14}{a} = \dfrac{70}{30}$

5) $\dfrac{5}{a} = \dfrac{15}{54}$

6) $\dfrac{\sqrt{4}}{5} = \dfrac{a}{60}$

7) $\dfrac{3}{9} = \dfrac{12}{a}$

8) $\dfrac{8}{16} = \dfrac{a}{31}$

9) $\dfrac{8}{a} = \dfrac{5.6}{7}$

10) $\dfrac{3}{16} = \dfrac{9}{a}$

11) $\dfrac{18}{9} = \dfrac{6}{a}$

12) $\dfrac{15}{a} = \dfrac{5}{25}$

13) $\dfrac{6}{7} = \dfrac{a}{13}$

14) $\dfrac{\sqrt{49}}{4} = \dfrac{42}{a}$

15) $\dfrac{7}{a} = \dfrac{7.7}{46.2}$

16) $\dfrac{80}{170} = \dfrac{a}{340}$

17) $\dfrac{32}{100} = \dfrac{a}{55}$

18) $\dfrac{43}{129} = \dfrac{a}{3}$

19) $\dfrac{20}{18} = \dfrac{3}{a}$

20) $\dfrac{8}{9} = \dfrac{64}{a}$

Reduce Ratio

Reduce each ratio to the simplest form.

1) 6: 24 =

2) 7: 42 =

3) 72: 40 =

4) 30: 25 =

5) 11: 110 =

6) 24: 3 =

7) 60: 300 =

8) 3: 108 =

9) 15: 45 =

10) 5.4: 6.3 =

11) 220: 660 =

12) 6: 10 =

13) 150: 250 =

14) 32: 48 =

15) 42: 84 =

16) 36: 9 =

17) 180: 45 =

18) 30: 300 =

19) 84: 56 =

20) 104: 132 =

21) 45: 90 =

22) 48: 56 =

23) 4: 60 =

24) 16: 96 =

Word Problems

Solve each word problem.

1) Bob has 18 red cards and 45 green cards. What is the ratio of Bob's red cards to his green cards? _____

2) In a party, 15 soft drinks are required for every 18 guests. If there are 378 guests, how many soft drinks is required? _____

3) In Bob's class, 45 of the students are tall and 25 are short. In Mason's class 117 students are tall and 65 students are short. Which class has a higher ratio of tall to short students? _____

4) The price of 3 apples at the Quick Market is $1.44. The price of 11 of the same apples at Walmart is $5.50. Which place is the better buy? _____

5) The bakers at a Bakery can make 300 bagels in 6 hours. How many bagels can they bake in 15 hours? What is that rate per hour? _____

6) You can buy 18 cans of green beans at a supermarket for $11.30. How much does it cost to buy 54 cans of green beans? _____

7) The ratio of boys to girls in a class is 7:5. If there are 21 boys in the class, how many girls are in that class? _____

8) The ratio of red marbles to blue marbles in a bag is 3:7. If there are 60 marbles in the bag, how many of the marbles are red? _____

Percent

Find the Percent of Numbers.

1) 20% of 40 =

2) 24% of 18 =

3) 12% of 15 =

4) 14% of 50 =

5) 7% of 70 =

6) 25% of 22 =

7) 15% of 45 =

8) 10% of 56 =

9) 30% of 74 =

10) 2.5% of 60 =

11) 75% of 16 =

12) 40% of 70 =

13) 14% of 120 =

14) 3% of 200 =

15) 35% of 0 =

16) 20% of 110 =

17) 32% of 55 =

18) 30% of 80 =

19) 9% of 14 =

20) 7% of 12 =

21) 18% of 46 =

22) 70% of 15 =

23) 12% of 55 =

24) 6% of 130 =

25) 80% of 260 =

26) 3% of 7 =

27) 11% of 420 =

28) 20% of 56 =

Convert Fraction to Percent

Write each as a percent.

1) $\dfrac{1}{5} =$

2) $\dfrac{2}{5} =$

3) $\dfrac{8}{16} =$

4) $\dfrac{3}{7} =$

5) $\dfrac{6}{14} =$

6) $\dfrac{13}{52} =$

7) $\dfrac{16}{22} =$

8) $\dfrac{21}{45} =$

9) $\dfrac{9}{75} =$

10) $\dfrac{3}{12} =$

11) $\dfrac{8}{24} =$

12) $\dfrac{36}{10} =$

13) $\dfrac{12}{40} =$

14) $\dfrac{32}{50} =$

15) $\dfrac{8}{29} =$

16) $\dfrac{3}{33} =$

17) $\dfrac{16}{44} =$

18) $\dfrac{14}{24} =$

19) $\dfrac{16}{86} =$

20) $\dfrac{9}{90} =$

21) $\dfrac{12}{300} =$

22) $\dfrac{560}{280} =$

Convert Decimal to Percent

Write each as a percent.

1) 0.159 =

2) 0.24 =

3) 3.1 =

4) 0.042 =

5) 0.007 =

6) 0.549 =

7) 0.157 =

8) 0.69 =

9) 0.004 =

10) 0.311 =

11) 0.782 =

12) 25.3 =

13) 7.585 =

14) 0.3 =

15) 4.82 =

16) 0.0254 =

17) 0.0079 =

18) 0.51 =

19) 4.15 =

20) 0.531 =

21) 6.323 =

22) 0.154 =

23) 3.9 =

24) 0.6 =

25) 1.7 =

26) 57.4 =

27) 3.05 =

28) 0.002 =

Answer key Chapter 7

Proportions

1) $a = 21$
2) $a = 4$
3) $a = 1$
4) $a = 6$
5) $a = 18$
6) $a = 24$
7) $a = 36$
8) $a = 15.5$
9) $a = 10$
10) $a = 48$
11) $a = 3$
12) $a = 75$
13) $a = \frac{78}{7}$
14) $a = 24$
15) $a = 42$
16) $a = 160$
17) $a = 17.6$
18) $a = 1$
19) $a = 2.7$
20) $a = 72$

Reduce Ratio

1) 1: 4
2) 1: 6
3) 9: 5
4) 6: 5
5) 1: 10
6) 8: 1
7) 1: 5
8) 1: 36
9) 1: 3
10) 0.6: 0.7
11) 11: 33
12) 0.6: 1
13) 3: 5
14) 2: 3
15) 1: 2
16) 4: 1
17) 4: 1
18) 1: 10
19) 3: 2
20) 26: 33
21) 1: 2
22) 6: 7
23) 1: 15
24) 1: 6

Word Problems

1) 2: 5
2) 315
3) The ratio for both classes is 9 to 5.
4) Quick Market is a better buy.
5) 750, the rate is 50per hour.
6) $33.90
7) 15
8) 18

Percent

1) 8
2) 4.32
3) 1.8
4) 7
5) 4.9
6) 5.5
7) 6.75
8) 5.6
9) 22.2
10) 1.5
11) 12
12) 28

13) 16.8
14) 6
15) 0
16) 22
17) 17.6
18) 24
19) 1.26
20) 0.84
21) 8.28
22) 10.5
23) 6.6
24) 7.8
25) 208
26) 0.21
27) 46.2
28) 11.2

Convert Fraction to Percent

1) 20%
2) 40%
3) 50%
4) 42.86%
5) 29.31%
6) 25%
7) 72.73%
8) 46.67%
9) 12%
10) 25%
11) 33.33%
12) 360%
13) 30%
14) 64%
15) 27.59%
16) 9.09%
17) 36.36%
18) 58.33%
19) 18.6%
20) 10%
21) 4%
22) 200%

Convert Decimal to Percent

1) 15.9%
2) 24%
3) 310%
4) 4.2%
5) 0.7%
6) 54.9%
7) 15.7%
8) 69%
9) 0.4%
10) 31.1%
11) 78.2%
12) 2,530%
13) 758.5%
14) 30%
15) 482%
16) 2.54%
17) 0.79%
18) 51%
19) 415%
20) 53.1%
21) 632.3%
22) 15.4%
23) 390%
24) 60%
25) 170%
26) 5,740%
27) 305%
28) 0.2%

Chapter 8:

Measurement

Reference Measurement

LENGTH	
Customary	**Metric**
1 mile (mi) = 1,760 yards (yd)	1 kilometer (km) = 1,000 meters (m)
1 yard (yd) = 3 feet (ft)	1 meter (m) = 100 centimeters (cm)
1 foot (ft) = 12 inches (in.)	1 centimeter(cm) = 10 millimeters(mm)
VOLUME AND CAPACITY	
Customary	**Metric**
1 gallon (gal) = 4 quarts (qt)	1 liter (L) = 1,000 milliliters (mL)
1 quart (qt) = 2 pints (pt.)	
1 pint (pt.) = 2 cups (c)	
1 cup (c) = 8 fluid ounces (Fl oz)	
WEIGHT AND MASS	
Customary	**Metric**
1 ton (T) = 2,000 pounds (lb.)	1 kilogram (kg) = 1,000 grams (g)
1 pound (lb.) = 16 ounces (oz)	1 gram (g) = 1,000 milligrams (mg)
Time	
1 year = 12 months	
1 year = 52 weeks	
1 week = 7 days	
1 day = 24 hours	
1 hour = 60 minutes	
1 minute = 60 seconds	

Metric Length Measurement

Convert to the units.

1) 2,000 mm = _____ cm

2) 5 m = _____ mm

3) 7 m = _____ cm

4) 9 km = _____ m

5) 5,000 mm = _____ m

6) 2,800 cm = _____ m

7) 13 m = _____ cm

8) 4,000 mm = _____ cm

9) 20,000 mm = _____ m

10) 7 km = _____ mm

11) 6 km = _____ m

12) 3 m = _____ cm

13) 17,000 m = _____ km

14) 500,000 m = _____ km

Customary Length Measurement

Convert to the units.

1) 15 ft = _____ in

2) 8 ft = _____ in

3) 7 yd = _____ ft

4) 9 yd = _____ ft

5) 3 yd = _____ in

6) 3 mi = _____ in

7) 7,200 in = _____ yd

8) 252 in = _____ yd

9) 8,800 yd = _____ mi

10) 12 yd = _____ in

11) 4 mi = _____ yd

12) 47,520 ft = _____ mi

13) 60 in = _____ ft

14) 25 yd = _____ ft

15) 36 in = _____ ft

16) 2 mi = _____ ft

Metric Capacity Measurement

Convert the following measurements.

1) 50 l = _____ ml

2) 4 l = _____ ml

3) 13 l = _____ ml

4) 8 l = _____ ml

5) 19 l = _____ ml

6) 2 l = _____ ml

7) 70,000 ml = _____ l

8) 8,000 ml = _____ l

9) 37,000 ml = _____ l

10) 200,000 ml = _____ l

11) 6,000,000 ml = _____ l

12) 40,000 ml = _____ l

Customary Capacity Measurement

Convert the following measurements.

1) 2 gal = _____ qt.

2) 11 gal = _____ pt.

3) 3 gal = _____ c.

4) 14 pt. = _____ c

5) 43 c = _____ fl oz

6) 16 qt = _____ pt.

7) 8 qt = _____ c

8) 29 pt. = _____ c

9) 6,720 c = _____ gal

10) 144 pt. = _____ gal

11) 72 qt = _____ gal

12) 92 pt. = _____ qt

13) 4,600 c = _____ qt

14) 146 c = _____ pt.

15) 108 qt = _____ gal

16) 1,848 pt. = _____ qt

17) 31 gal = _____ pt.

18) 6 qt = _____ c

19) 640 c = _____ gal

20) 104 fl oz = _____ c

Metric Weight and Mass Measurement

Convert.

1) 7 kg = _____ g

2) 3 kg = _____ g

3) 13 kg = _____ g

4) 21 kg = _____ g

5) 9 kg = _____ g

6) 121 kg = _____ g

7) 249 kg = _____ g

8) 4,000 g = _____ kg

9) 6,000 g = _____ kg

10) 17,000 g = _____ kg

11) 129,000 g = _____ kg

12) 220,000 g = _____ kg

13) 9,000,000 g = _____ kg

14) 11,000,000 g = _____ kg

Customary Weight and Mass Measurement

Convert.

1) 16,000 lb. = _____ T

2) 20,000 lb. = _____ T

3) 170,000 lb. = _____ T

4) 44,000 lb. = _____ T

5) 7 lb. = _____ oz

6) 4 lb. = _____ oz

7) 10 lb. = _____ oz

8) 24 T = _____ lb.

9) 3 T = _____ lb.

10) 9 T = _____ lb.

11) 112 T = _____ lb.

12) 2 T = _____ oz

13) 5 T = _____ oz

14) 224 oz = _____ lb.

Time

Convert to the units.

1) 16 hr. = _____ min

2) 9 year = _____ week

3) 2 hr. = _____ sec

4) 12 min = _____ sec

5) 600 min = _____ hr

6) 730 day = _____ year

7) 3 year = _____ hr.

8) 27 day = _____ hr

9) 4 day = _____ min

10) 540 min = _____ hr

11) 16 year = _____ month

12) 19,200 sec = _____ min

13) 288 hr = _____ day

14) 23 weeks = _____ day

How much time has passed?

15) From 4:15 A.M. to 7:25 A.M.: ____ hours and ____ minutes.

16) From 3:40 A.M. to 8:25 A.M.: ____ hours and ____ minutes.

17) It's 8:50 P.M. What time was 2 hours ago? _____ O'clock

18) 3:20 A.M to 6:40 AM: _____ hours and _____ minutes.

19) 3:30 A.M to 6:05 AM: _____ hours and _____ minutes.

20) 7:10 A.M. to 8:15 AM. = _____ hour(s) and _____ minutes.

21) 11:55 A.M. to 4:25 PM. = _____ hour(s) and _____ minutes

22) 7:18 A.M. to 7:52 A.M. = _____ minutes

23) 9:13 A.M. to 9:50 A.M. = _____ minutes

Answers of Worksheets – Chapter 8

Metric length

1) 200 cm
2) 5,000 mm
3) 700 cm
4) 9,000 m
5) 5 m
6) 28 m
7) 1,300 cm
8) 400 cm
9) 20 m
10) 7,000,000 mm
11) 6,000 m
12) 300 cm
13) 17 km
14) 500 km

Customary Length

1) 180
2) 96
3) 21
4) 27
5) 108
6) 190,080
7) 200
8) 7
9) 5
10) 432
11) 7,040
12) 9
13) 5
14) 75
15) 3
16) 10,560

Metric Capacity

1) 50,000 ml
2) 4,000 ml
3) 13,000 ml
4) 8,000 ml
5) 19,000 ml
6) 2,000 ml
7) 70 L
8) 8 L
9) 37 L
10) 200 L
11) 6,000 L
12) 40 L

Customary Capacity

1) 8 qt
2) 88 pt.
3) 48 c
4) 28 c
5) 344 fl oz
6) 12 pt.
7) 32 c
8) 58 c
9) 420 gal
10) 18 gal
11) 18 gal
12) 46 qt
13) 1,150 qt
14) 73 pt.
15) 27 gal
16) 927 qt
17) 248 pt.
18) 24 c
19) 40 gal
20) 13 c

Metric Weight and Mass

1) 7,000 g
2) 3,000 g
3) 13,000 g
4) 21,000 g
5) 9,000 g
6) 121,000 g
7) 249,000 g
8) 4 kg
9) 6 kg
10) 17 kg
11) 129 kg
12) 220 kg
13) 9,000 kg
14) 11,000 kg

Customary Weight and Mass

1) 8 T
2) 10 T
3) 85 T
4) 22 T
5) 112 oz
6) 64 oz
7) 160 oz
8) 48,000 lb.
9) 6,000 lb.
10) 18,000 lb.
11) 224,000 lb.
12) 64,000 oz
13) 160,000 oz
14) 14 lb

Time

1) 960 min
2) 468 weeks
3) 7,200 sec
4) 720 sec
5) 10 hr
6) 2 year
7) 26,280 hr
8) 648 hr
9) 5,760 min
10) 9 hr
11) 192 months
12) 320 min
13) 12 days
14) 161 days
15) 3:10
16) 4:45
17) 6:50 P.M.
18) 3:20
19) 2:35
20) 1:05
21) 4:30
22) 34 minutes
23) 37 minutes

Chapter 9: Algebraic Expressions

Find a Rule

Complete the output.

1- **Rule: the output is** $x + 15$

Input	x	7	11	18	26	34
Output	y					

2- **Rule: the output is** $x \times 12$

Input	x	2	5	9	13	17
Output	y					

3- **Rule: the output is** $x \div 8$

Input	x	104	184	136	312	440
Output	y					

Find a rule to write an expression.

4- **Rule:** _____

Input	x	12	14	17	21
Output	y	72	84	102	126

5- **Rule:** _____

Input	x	9	18	27	39
Output	y	17	26	33	47

6- **Rule:** _____

Input	x	112	168	224	357
Output	y	16	24	32	51

Variables and Expressions

Write a verbal expression for each algebraic expression.

1) $3a - 7b$

2) $5c^2 + 9d$

3) $x - 13$

4) $\frac{90}{17}$

5) $m^2 + n^3$

6) $6x + 3$

7) $a^2 - 11b + 6$

8) $x^3 + 8y^2 - 6$

9) $\frac{1}{6}x + \frac{3}{4}y - 9$

10) $\frac{1}{3}(x + 7) - 12y$

Write an algebraic expression for each verbal expression.

11) 5 less than y

12) The product of 6 and a

13) The 32 divided by z

14) The product of 4 and the third power of a

15) 6 more than h to the seventh power

16) 43 more than triple d

17) One fifth the square of x

18) The difference of 34 and 6 times a number

19) 57 more than the cube of a number

20) One-quarters the cube of a number

Translate Phrases

Write an algebraic expression for each phrase.

1) A number increased by fifty–four.

2) The sum of ninety and 4 times a number

3) The difference between seventy–eight and a number.

4) The quotient of sixty-one and a number.

5) Triple a number decreased by 80.

6) two times the sum of a number and – 45.

7) A number divided by – 18.

8) The quotient of 38 and the product of a number and – 12.

9) nine subtracted from 4 times a number.

10) The difference of six and a number.

Distributive Property

Multiply using the distributive property.

1) $3(x + 3) =$ _____

2) $4(x + 11) =$ _____

3) $(x + 9)5 =$ _____

4) $7(x + 6) =$ _____

5) $8(x + 8) =$ _____

6) $10(x + 4) =$ _____

7) $9(x + 10) =$ _____

8) $4(x + 8) =$ _____

9) $11(x + 6) =$ _____

10) $(x + 7)6 =$ _____

11) $(x + 13)4 =$ _____

12) $3(x + 12) =$ _____

13) $2(9x - 4) =$ _____

14) $7(8x - 3) =$ _____

15) $8(9x - 5) =$ _____

16) $(4x - 2)3 =$ _____

17) $(7x - 2)7 =$ _____

18) $(2x - 3)12 =$ _____

19) $3(4x - 1) =$ _____

20) $(-2)(4x - 3) =$ _____

21) $(-5)(x - 9) =$ _____

22) $(-7)(3x - 1) =$ _____

23) $(5x + 2)(-9) =$ _____

24) $(x + 6)(-12) =$ _____

Evaluate One Variable Expressions

Evaluate each using the values given.

1) $x + 2x, x = 5$

2) $3(-7 + x), x = 2$

3) $6x + 4x, x = -1$

4) $2(6 - x) + 8, x = 4$

5) $5x + 4x - 15, x = 3$

6) $3x + 7x + 22, x = -2$

7) $4x - 3x - 6, x = 4$

8) $\frac{2(2x+10)}{4}, x = 3$

9) $3x - 87, x = 25$

10) $\frac{x}{23}, x = 138$

11) $3(4 + 3x) - 24, x = 3$

12) $8(x + 2) - 31, x = 5$

13) $\frac{x+(-7)}{-2}, x = -7$

14) $6(5 - 2x) + 8, x = 3$

15) $-12 - \frac{x}{4} + 6x, x = 20$

16) $3x + 13x, x = 2$

17) $-10x + 2(6 + 2x), x = -5$

18) $3x + 15x, x = 3$

19) $\frac{(3x-7)}{5}, x = 14$

20) $4(-2 - 3x), x = 7$

21) $6x - (6 - x), x = 2$

22) $\left(-\frac{30}{x}\right) + 3 + x, x = 6$

23) $-\frac{x \times 8}{x}, x = 8$

24) $3(-2 - 4x), x = 3$

25) $3x^2 + 5x, x = 2$

26) $7(4x + 2) - 3(x - 8), x = 1$

27) $-4x - 3, x = -6$

28) $6x + 3x, x = 2$

Answer key Chapter 9

Find a Rule

1)
Input	x	7	11	18	26	34
Output	y	22	26	33	41	49

2)
Input	x	2	5	9	13	17
Output	y	24	60	108	156	204

3)
Input	x	104	184	136	312	440
Output	y	13	23	17	39	55

4) $y = 6x$ 5) $y = x + 8$ 6) $y = x \div 7$

Variables and Expressions

1) 3 times a minus 7 times b

2) 5 times c squared plus 9 times d

3) a number minus 13

4) the quotient of 90 and 17

5) m squared plus n cubed

6) the product of 6 and x plus 3

7) a squared minus the product of 11 and b plus 6

8) x cubed plus the product of 8 and y squared minus 6

9) the sum of one–sixth of x and three–quarters of y, minus 9

10) one–fifth of the sum of x and 7 minus the product of 12 and y

11) 5 <y

12) 6a

13) $\frac{32}{z}$

14) $4a^3$

15) $6 > h^7$

16) $3d < 43$

17) $\frac{1}{5}x^2$

18) $34 - 6a$

19) $57 > a^3$

20) $\frac{1}{4}x^3$

Translate Phrases

1) $x + 54$

2) $90 + 4x$

3) $78 - x$

4) $\frac{61}{x}$

5) $3x - 80$

6) $2(x + (-45))$

7) $\frac{x}{-18}$

8) $\frac{38}{-12x}$

9) $4x - 9$

10) $6 - x$

Distributive Property

1) $3x + 9$

2) $4x + 44$

3) $5x + 45$

4) $7x + 42$

5) $8x + 64$

6) $10x + 40$

WWW.MathNotion.com

SBAC Math Practice Grade 5

7) $9x + 90$
8) $4x + 32$
9) $11x + 66$
10) $6x + 42$
11) $4x + 52$
12) $3x + 36$

13) $18x - 8$
14) $56x - 21$
15) $72x - 40$
16) $12x - 6$
17) $49x - 14$
18) $24x - 36$

19) $12x - 3$
20) $-8x + 6$
21) $-5x + 45$
22) $-21x + 7$
23) $-45x - 18$
24) $-12x - 72$

Evaluate One Variable Expressions

1) 15
2) −15
3) −10
4) 12
5) 12
6) 2
7) −2

8) 8
9) −12
10) 6
11) 15
12) 25
13) 7
14) 2

15) 103
16) 32
17) 42
18) 54
19) 7
20) −92
21) 8

22) 4
23) −8
24) −42
25) 22
26) 63
27) 21
28) 18

Chapter 10: Symmetry and Transformations

Line Segments

Write each as a line, ray, or line segment.

1)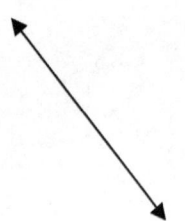

2)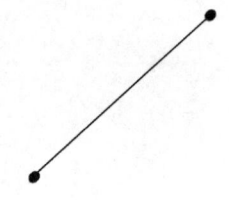

3)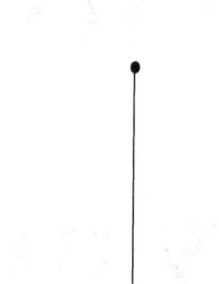

4)

5)

6)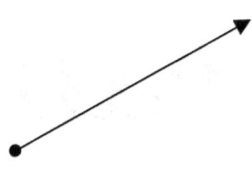

7)

8)

Parallel, Perpendicular and Intersecting Lines

State whether the given pair of lines are parallel, perpendicular, or intersecting.

1)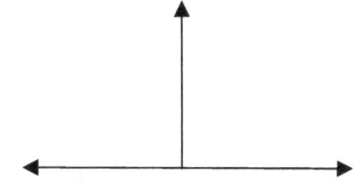

2)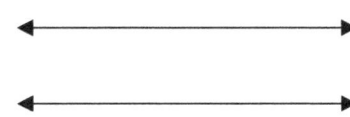

3)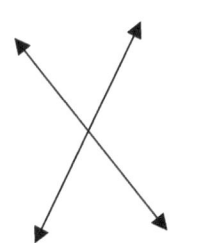

4)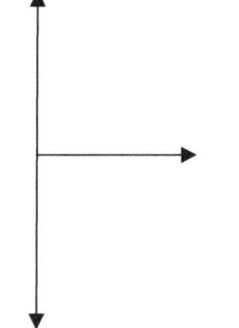

5)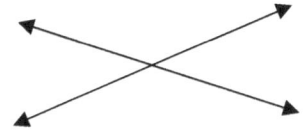

6)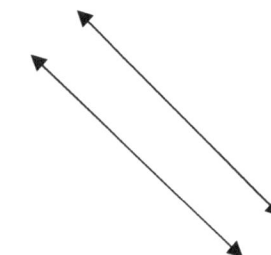

7)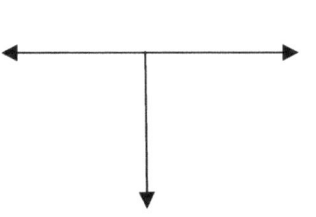

8)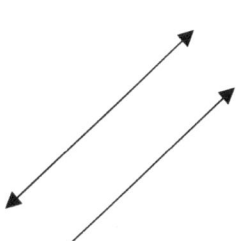

Identify Lines of Symmetry

Tell whether the line on each shape a line of symmetry is.

1)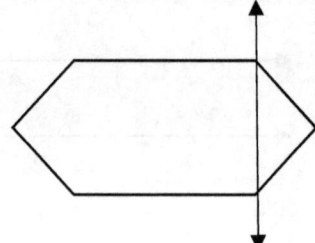

2)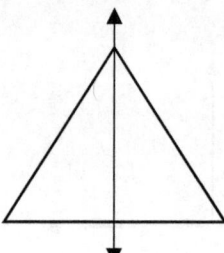

3)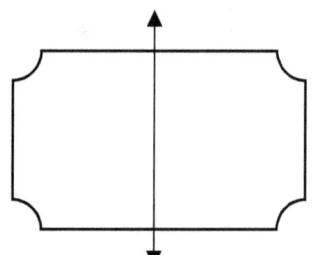

4)

5)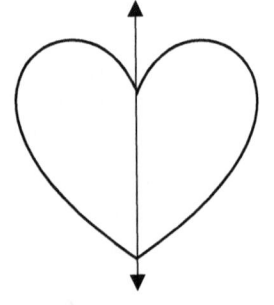

6)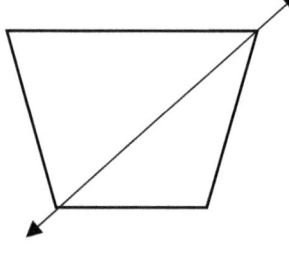

7)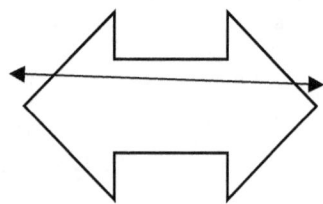

8)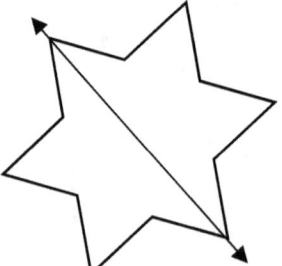

Lines of Symmetry

Draw lines of symmetry on each shape. Count and write the lines of symmetry you see.

1)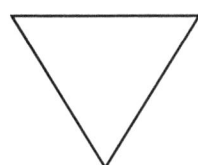

2)

3)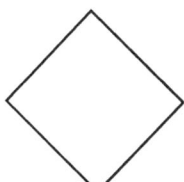

4)

5)

6)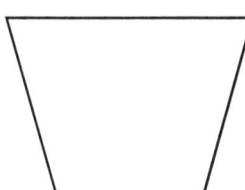

7)

8)

Identify Three–Dimensional Figures

Write the name of each shape.

1)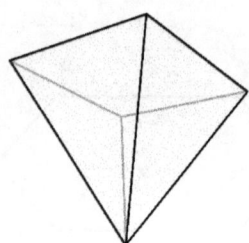

2)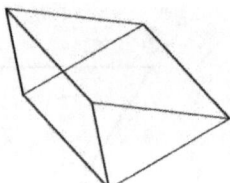

3)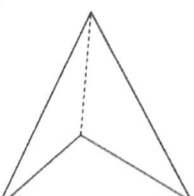

4)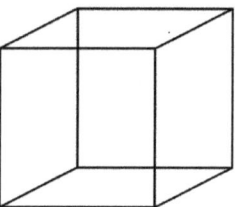

5)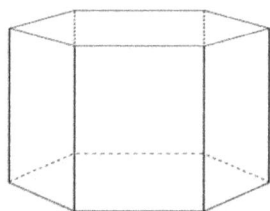

6)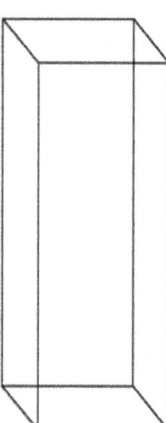

7)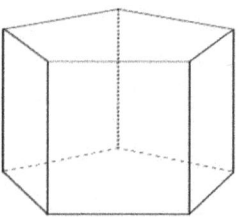

Vertices, Edges, and Faces

Complete the chart below.

Shape	Number of edges	Number of faces	Number of vertices
1) hexagonal prism	_____	_____	_____
2) rectangular prism	_____	_____	_____
3) tetrahedron	_____	_____	_____
4) square pyramid	_____	_____	_____
5) cube	_____	_____	_____
6) rectangular prism	_____	_____	_____

Identify Faces of Three–Dimensional Figures

Write the number of faces.

1)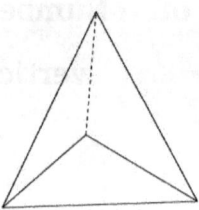

2)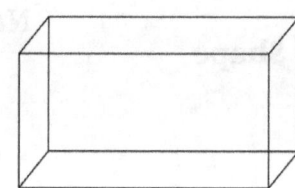

3)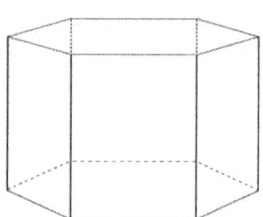

4)

5)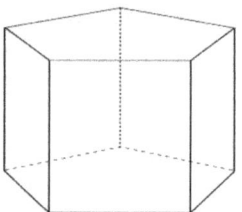

6)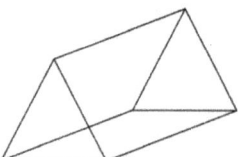

7)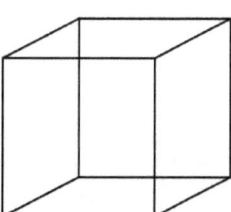

8)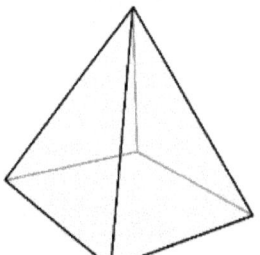

Answers of Worksheets – Chapter 10

Line Segments

1) Line
2) Line segment
3) Line segment
4) Ray
5) Line
6) Ray
7) Line segment
8) Ray

Parallel, Perpendicular and Intersecting Lines

1) Perpendicular
2) Parallel
3) Intersection
4) Perpendicular
5) Intersection
6) Parallel
7) Perpendicular
8) Parallel

Identify lines of symmetry

1) No
2) yes
3) yes
4) No
5) yes
6) No
7) No
8) yes

lines of symmetry

1)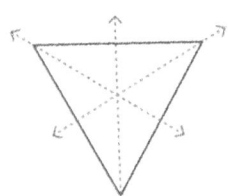

2)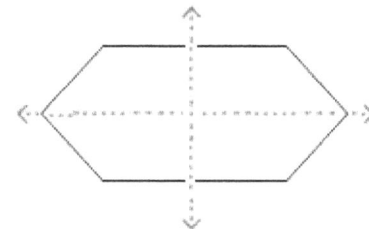

3)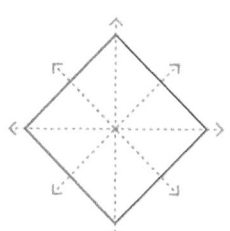

4)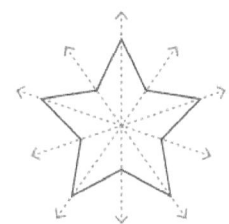

SBAC Math Practice Grade 5

5)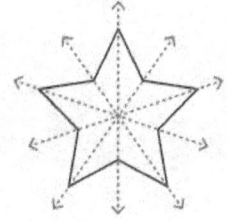

6)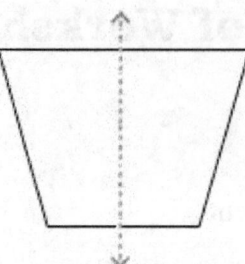

7)

8)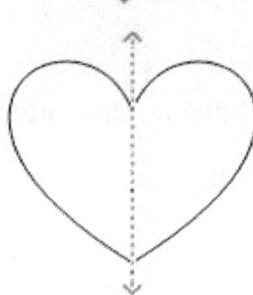

Identify Three–Dimensional Figures

1) Square pyramid
2) Triangular prism
3) Triangular pyramid
4) Cube
5) Hexagonal prism
6) Rectangular prism
7) Pentagonal prism

Vertices, Edges, and Faces

Shape	Number of edges	Number of faces	Number of vertices
1)	18	8	12
2)	12	6	8
3)	6	4	4

SBAC Math Practice Grade 5

4) 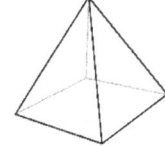   8   5   5

5) 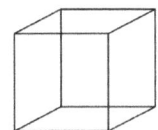   12   6   8

6) 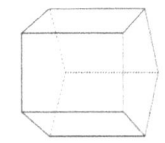   15   7   10

Identify Faces of Three–Dimensional Figures

1) 4   4) 2   7) 6
2) 6   5) 7   8) 5
3) 8   6) 5

Chapter 11:
Geometry

Area and Perimeter of Square

Find the perimeter and area of each squares.

1)

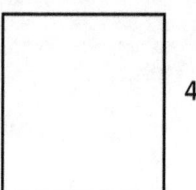

Perimeter:_____:

Area:_____:

2)

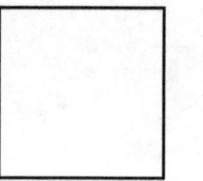

Perimeter:_____:

Area:_____:

3)

Perimeter:_____:

Area:_____:

4)

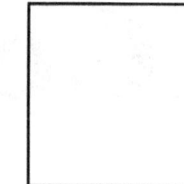

Perimeter:_____:

Area:_____:

5)

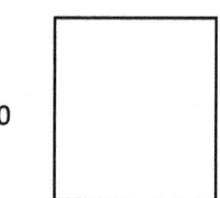

Perimeter:_____:

Area:_____:

6)

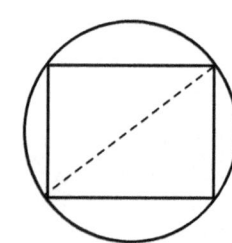

Perimeter of Square:_____:

Area of Square:_____:

Area and Perimeter of Rectangle

Find the perimeter and area of each rectangle.

1)

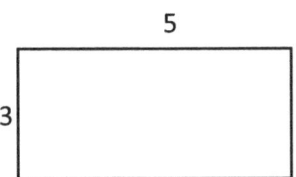

Perimeter:_____:

Area:_____:

2)

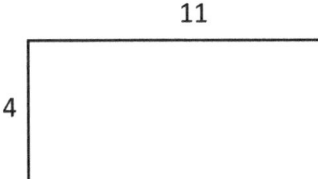

Perimeter:_____:

Area:_____:

3)

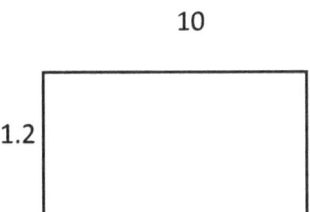

Perimeter:_____:

Area:_____:

4)

$9\frac{1}{2}$

2

Perimeter:_____:

Area:_____:

5)

7.2

5

Perimeter:_____:

Area:_____:

6)

8.4

3.6

Perimeter:_____:

Area:_____:

Area and Perimeter of Triangle

Find the perimeter and area of each triangle.

1)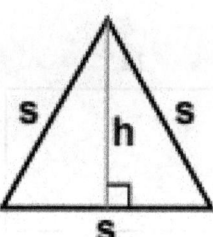

Perimeter:_____:

Area:_____:

2)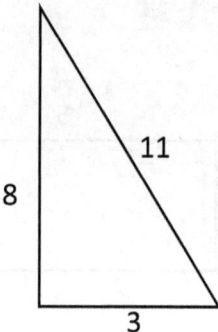

Perimeter:_____:

Area:_____:

3)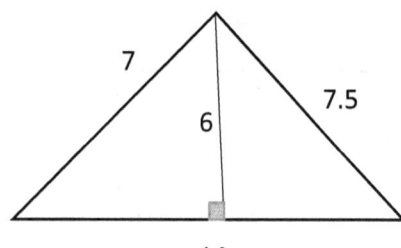

Perimeter:_____:

Area:_____:

4) s=10

h=6.4

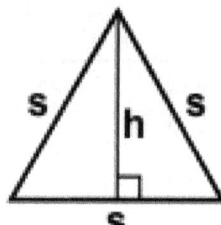

Perimeter:_____:

Area:_____:

5)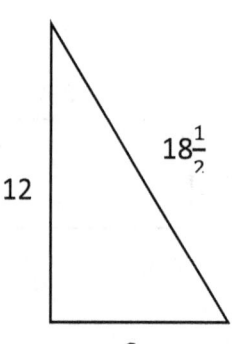

Perimeter:_____:

Area:_____:

6)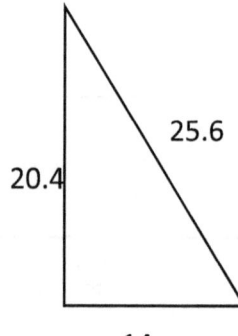

Perimeter:_____:

Area:_____:

Area and Perimeter of Trapezoid

Find the perimeter and area of each trapezoid.

1)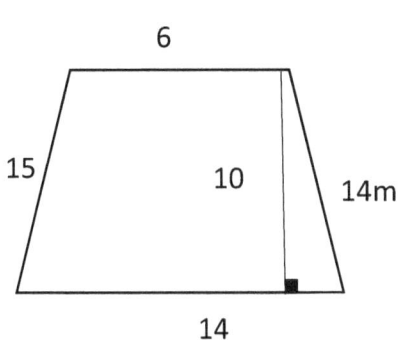

Perimeter:_____.

Area:_____.

2)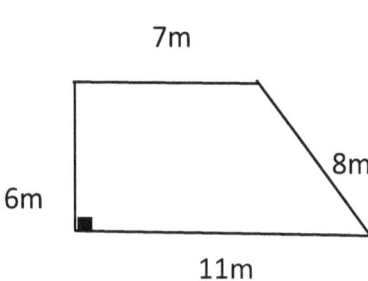

Perimeter:_____.

Area:_____.

3)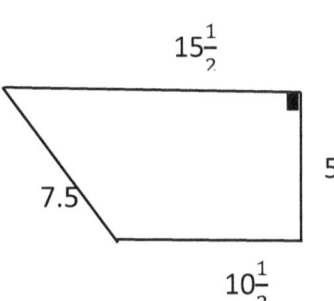

Perimeter:_____.

Area:_____.

4)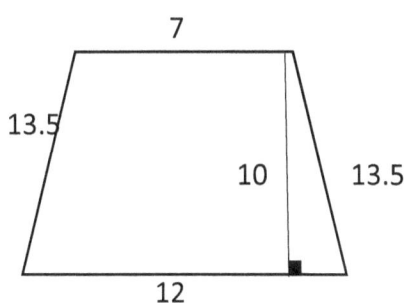

Perimeter:_____.

Area:_____.

5)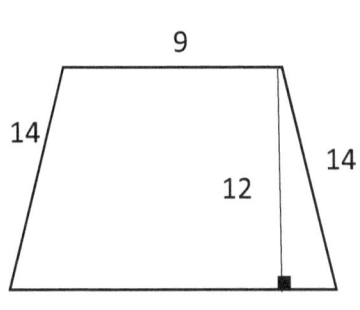

Perimeter:_____.

Area:_____.

6)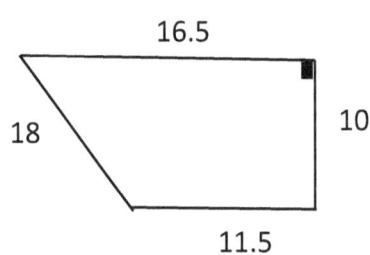

Perimeter:_____.

Area:_____.

Area and Perimeter of Parallelogram

Find the perimeter and area of each parallelogram.

1)

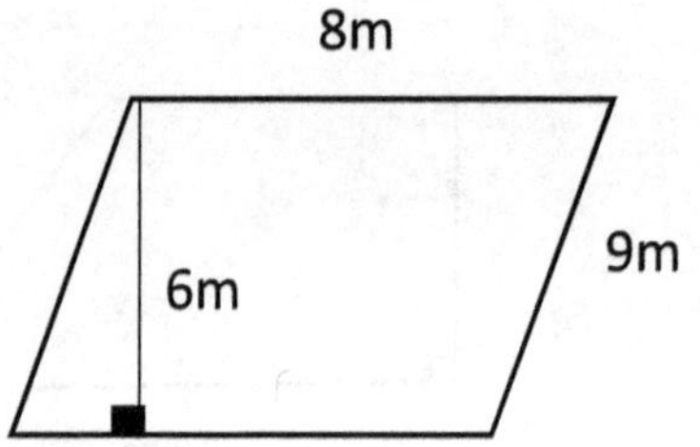

Perimeter:..............

Area:......................

2)

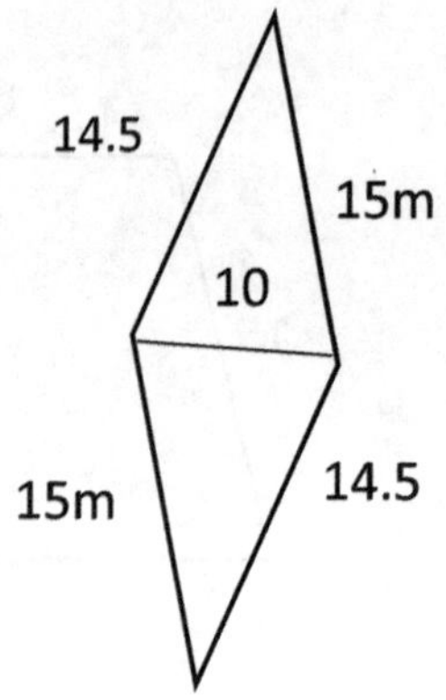

Perimeter:.....................

Area:......................

3)

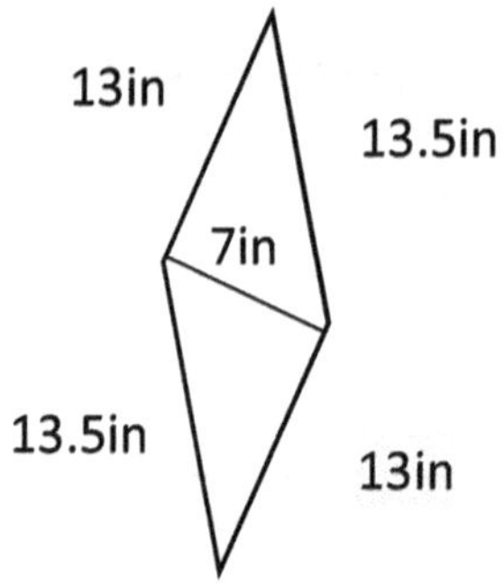

Perimeter:....................

Area:......................

4)

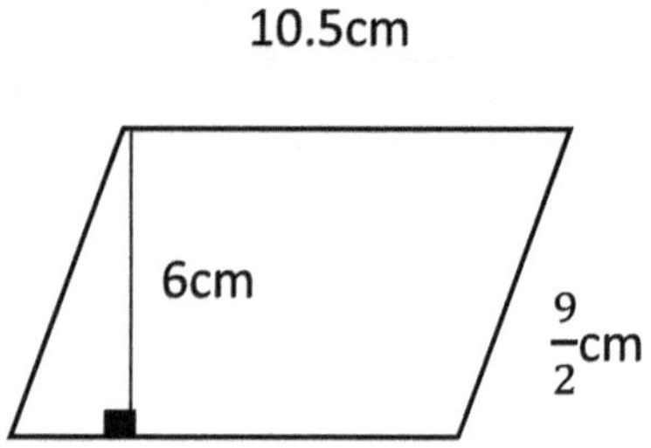

Perimeter:.......................

Area:.....................

5)

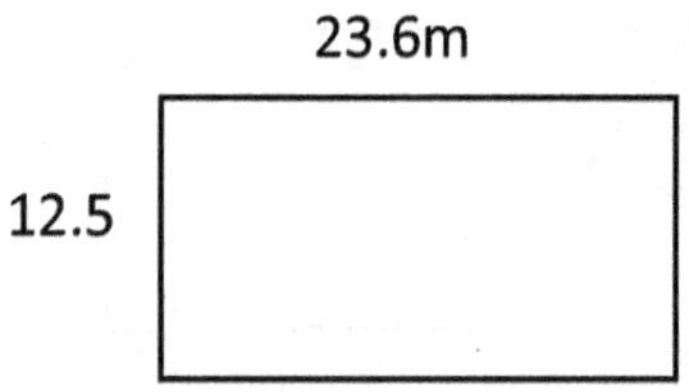

Perimeter:.........

Area:................

6)

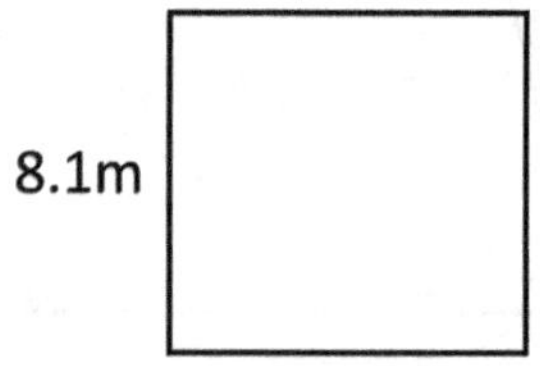

Perimeter:.....................

Area:.....................

Circumference and Area of Circle

Find the circumference and area of each ($\pi = 3.14$).

1)

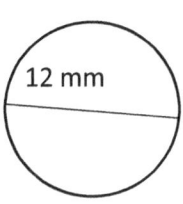

Circumference:

Area:

2)

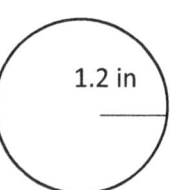

Circumference: _____

Area: _____

3)

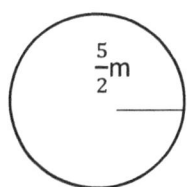

Circumference: _____

Area: _____

4)

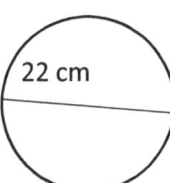

Circumference: _____

Area: _____

5)

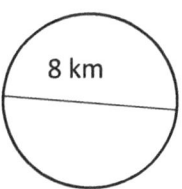

Circumference: _____

Area: _____

6)

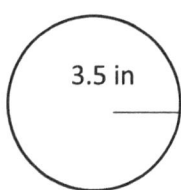

Circumference: _____

Area: _____

Perimeter of Polygon

Find the perimeter of each polygon.

1)

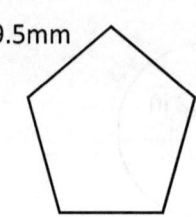
9.5mm

Perimeter:_____.

2)

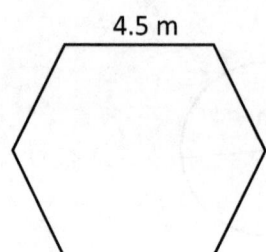
4.5 m

Perimeter:_____:

3)

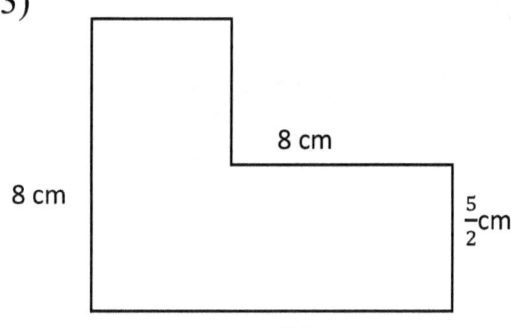
8 cm, 8 cm, $\frac{5}{2}$ cm, 12.5 cm

Perimeter:_____.

4)

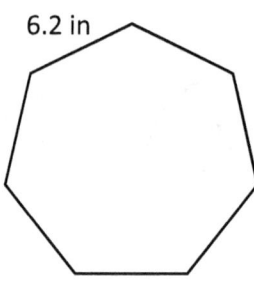
6.2 in

Perimeter:_____:

5)

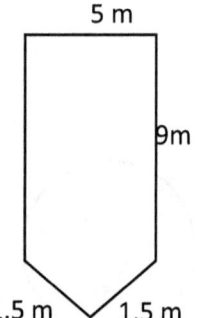
5 m, 9m, 1.5 m, 1.5 m

Perimeter:_____.

6)

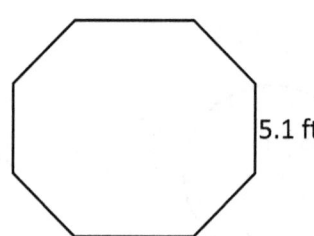
5.1 ft

Perimeter:_____:

Volume of Cubes

Find the volume of each cube.

1)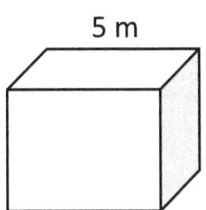

V:..

2)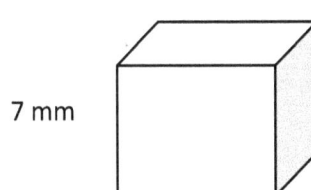

V:..

3)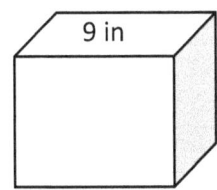

V:..

4)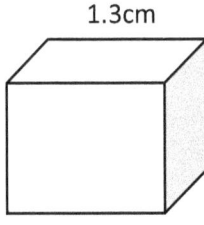

V:..

5)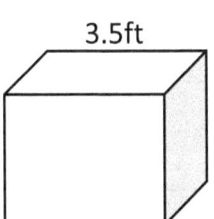

V:..

6)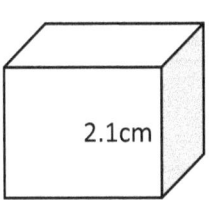

V:..

Volume of Rectangle Prism

Find the volume of each rectangle prism

1)

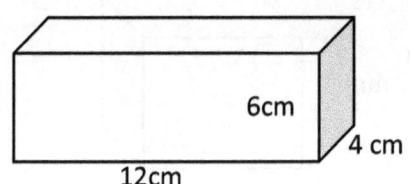

V:..

2)

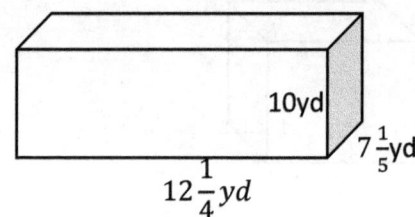

V:..

3)

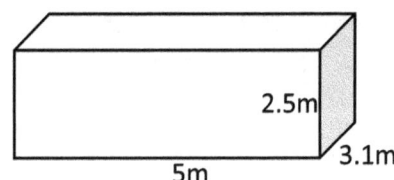

V:..

4)

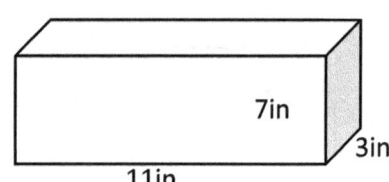

V:..

5)

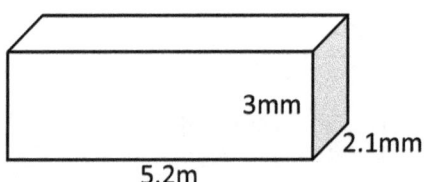

V:..

6)

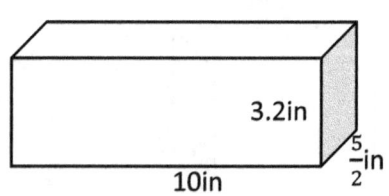

V:..

Answer key Chapter 11

Area and Perimeter of Square
1. Perimeter: 16,   Area: 16
2. Perimeter: 6,   Area: 2.25
3. Perimeter: 10, Area: 6.25
4. Perimeter: 32,   Area: 64
5. Perimeter: 40,   Area: 100
6. Perimeter: $4\sqrt{3}$, Area: 3

Area and Perimeter of Rectangle
1- Perimeter: 16,   Area: 15
2- Perimeter: 30,   Area: 44
3- Perimeter: 22.4, Area: 12
4- Perimeter: 23, Area: 19
5- Perimeter: 24.4, Area: 36
6- Perimeter: 24, Area: 30.24

Area and Perimeter of Triangle
1- Perimeter: 3s, Area: $\frac{1}{2}sh$
2- Perimeter: 22,   Area: 12
3- Perimeter: 28.5, Area: 42
4- Perimeter: 30,   Area: 32
5- Perimeter: 38.5, Area: 48
6- Perimeter: 60, Area: 142.8

Area and Perimeter of Trapezoid
1- Perimeter: 49,   Area: 100
2- Perimeter: 32,   Area: 54
3- Perimeter: 38.5, Area: 65
4- Perimeter: 46,   Area: 95
5- Perimeter: 52,   Area: 144
6- Perimeter: 56,   Area: 140

Area and Perimeter of Parallelogram
1- Perimeter: $34m$, Area: $48(m)^2$
2- Perimeter: $59m$, Area: $145(m)^2$
3- Perimeter: $53in$, Area: $91(in)^2$
4- Perimeter: $30cm$, Area: $63(cm)^2$
5- Perimeter: $72.2m$, Area: $295(m)^2$
6- Perimeter: $32.4m$, Area: $65.61(m)^2$

Circumference and Area of Circle
1) Circumference: 37.68 mm   Area: $113.04(mm)^2$
2) Circumference: 7.536 in   Area: $4.522(in)^2$
3) Circumference: 15.7 m   Area: $19.625(m)^2$
4) Circumference: 69.08 cm   Area: $379.94(cm)^2$
5) Circumference: 25.12 km   Area: $50.24(km)^2$
6) Circumference: 21.98 in   Area: $38.465(in)^2$

Perimeter of Polygon
1) 47.5 mm
2) 27 m
3) 41 cm
4) 43.4 in
5) 26 m
6) 40.8 ft

Volume of Cubes
1) $125m^3$
2) $343(mm)^3$
3) $729in^3$
4) $2.197(cm)^3$

5) $42.875 (ft)^3$ 6) $9.261 (cm)^3$

Volume of Rectangle Prism

1) $288 (cm)^3$ 3) $38.75 (m)^3$ 5) $32.76 (mm)^3$

2) $882 (yd)^3$ 4) $231 (in)^3$ 6) $80 (in)^3$

Chapter 12: Data and Graphs

Mean and Median

Find the mean and median of the following data.

1) 12, 49, 23, 48, 15, 29, 69

Mean: __, Median: __

2) 8, 16, 16, 29, 17, 40

Mean: __, Median: __

3) 23, 75, 55, 18, 71, 19, 23, 10, 8

Mean: __, Median: __

4) 50, 22, 2, 30, 50, 36

Mean: __, Median: __

5) 9, 11, 19, 22, 43, 27, 4, 2, 44, 7

Mean: __, Median: __

6) 5, 13, 17, 35, 10, 33, 1, 66

Mean: __, Median: __

7) 20, 12, 33, 6, 19, 81, 9, 34, 25, 55

Mean: __, Median: __

8) 22, 17, 46, 13, 22

Mean: __, Median: __

9) 32, 32, 51, 23, 69

Mean: __, Median: __

10) 22, 11, 7, 8, 10, 12, 14

Mean: __, Median: __

11) 12, 36, 12, 18, 3, 24

Mean: __, Median: __

12) 16, 34, 28, 28, 11, 12, 11

Mean: __, Median: __

13) 5, 5, 6, 10, 2, 11, 30

Mean: __, Median: __

14) 33, 11, 15, 22, 22, 18, 39

Mean: __, Median: __

15) 12, 23, 23, 18, 34

Mean: __, Median: __

16) 7, 7, 19, 14, 16, 28, 31

Mean: __, Median: __

17) 31, 10, 12, 28, 19, 17, 6, 2

Mean: __, Median: __

18) 35, 39, 24, 58, 65, 4

Mean: __, Median: __

19) 43, 38, 19, 58, 3, 7

Mean: __, Median: __

20) 57, 46, 37, 18, 12, 66, 14

Mean: __, Median: __

Mode and Range

Find the mode(s) and range of the following data.

1) 16, 73, 10, 46, 12, 16, 57

Mode: __, Range: __

2) 7, 18, 18, 21, 14, 16

Mode: __, Range: __

3) 25, 32, 28, 49, 27, 25, 18, 13

Mode: __, Range: __

4) 28, 19, 3, 14, 37, 14

Mode: __, Range: __

5) 31, 18, 18, 15, 43, 15, 18, 59

Mode: __, Range: __

6) 8, 16, 12, 19, 8, 12, 24, 12

Mode: __, Range: __

7) 34, 18, 55, 46, 41, 34, 18, 34, 14, 34

Mode: __, Range: __

8) 72, 34, 7, 72, 29, 51

Mode: __, Range: __

9) 18, 18, 38, 52, 49

Mode: __, Range: __

10) 16, 10, 1, 1, 6, 12, 19

Mode: __, Range: __

11) 8, 17, 11, 9, 4, 9

Mode: __, Range: __

12) 16, 8, 22, 19, 21, 21, 42

Mode: __, Range: __

13) 5, 5, 2, 17, 4, 5, 29

Mode: __, Range: __

14) 14, 16, 20, 14, 14, 17, 27, 31

Mode: __, Range: __

15) 7, 16, 14, 16, 35

Mode: __, Range: __

16) 8, 8, 12, 19, 39, 43, 48

Mode: __, Range: __

17) 31, 19, 12, 33, 17, 21, 19, 1

Mode: __, Range: __

18) 48, 51, 48, 27, 67, 6

Mode: __, Range: __

19) 43, 15, 15, 72, 51, 3

Mode: __, Range: __

20) 40, 67, 28, 27, 14, 49, 67

Mode: __, Range: __

Stem-And-Leaf Plot

Make stem-and-leaf plots for the given data.

1) 25, 26, 47, 40, 22, 65, 28, 46, 48, 43, 29, 41, 66

2) 51, 54, 19, 30, 52, 10, 36, 59, 38, 35, 39, 32

3) 114, 88, 45, 45, 85, 117, 41, 116, 49, 112, 46, 89

4) 71, 55, 76, 70, 106, 72, 101, 103, 79, 72, 59, 51

5) 63, 68, 137, 64, 27, 139, 133

6) 141, 37, 95, 32, 141, 148, 92, 146, 37, 97, 90

Dot plots

The ages of students in a Math class are given below.

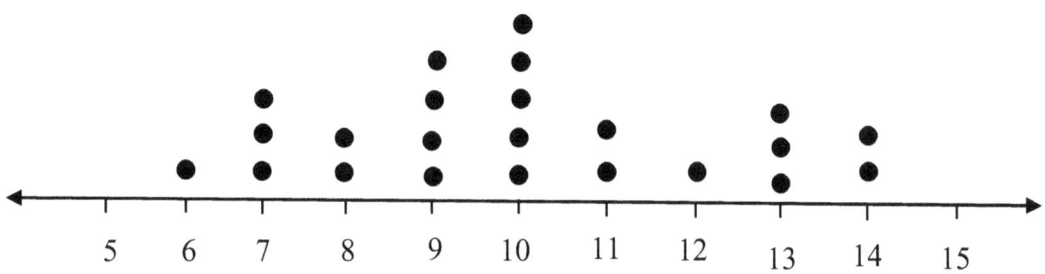

1) What is the total number of students in math class?

2) How many students are at least 13 years old?

3) Which age(s) has the most students?

4) Which age(s) has the fewest student?

5) Determine the median of the data.

6) Determine the range of the data.

7) Determine the mode of the data.

Bar Graph

Each student in class selected two games that they would like to play. Graph the given information as a bar graph and answer the questions below:

Game	Votes
Football	16
Volleyball	11
Basketball	19
Baseball	15
Tennis	11

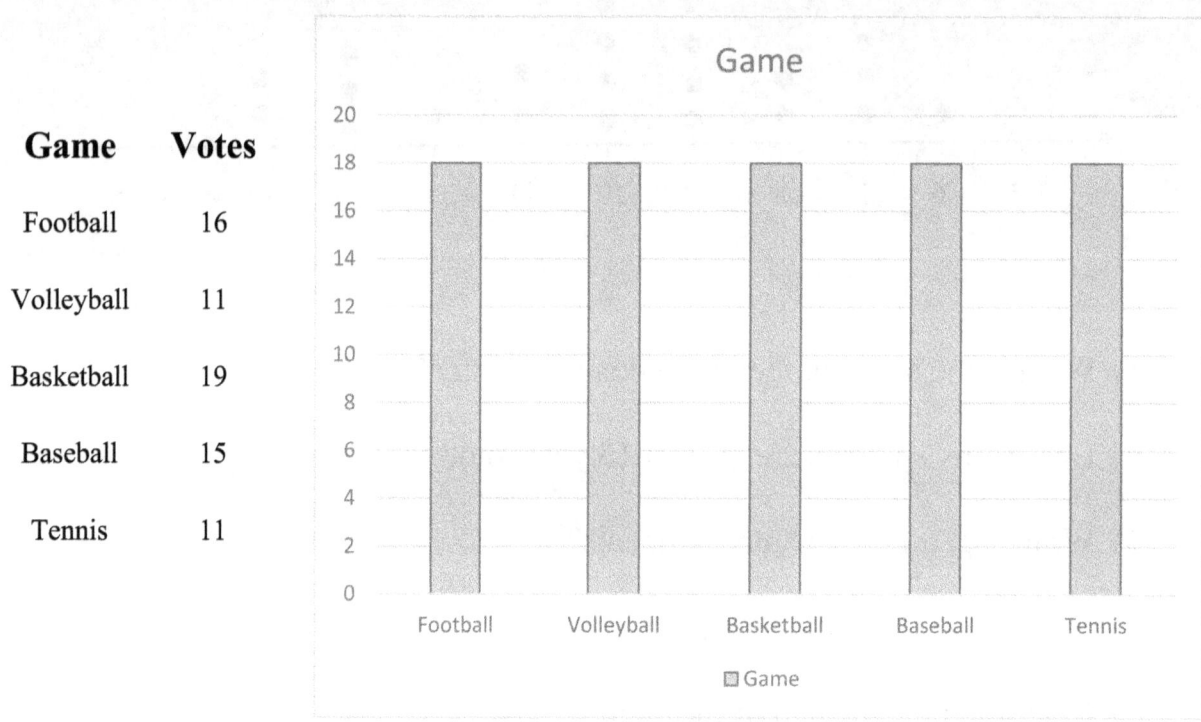

1) Which was the most popular game to play?

2) How many more students like Basketball than Volleyball?

3) Which two game got the same number of votes?

4) How many Volleyball and Football did students vote in all?

5) Did more students like football or Volleyball?

6) Which game(s) did the fewest students like?

Probability

1) A jar contains 15 caramels, 8 mints and 17 dark chocolates. What is the probability of selecting a mint?

2) If you were to roll the dice one time what is the probability it will NOT land on a 3?

3) A die has sides are numbered 1 to 6. If the cube is thrown once, what is the probability of rolling a 2?

4) The sides of number cube have the numbers 2, 4, 6, 2, 4, and 6. If the cube is thrown once, what is the probability of rolling a 4?

5) Your friend asks you to think of a number from one to ten. What is the probability that his number will be 7?

6) A person has 12 coins in their pocket. 3 dime, 5 pennies, 4 quarter, and a nickel. If a person randomly picks one coin out of their pocket. What would the probability be that they get a penny?

7) What is the probability of drawing an odd numbered card from a standard deck of shuffled cards (Ace is one)?

8) 40 students apply to go on a school trip. Three students are selected at random. what is the probability of selecting 5 students?

Answer key Chapter 12

Mean and Median

1) mean: 35, median: 29
2) mean: 21, median: 16.5
3) mean: 33.56, median: 23
4) mean: 31.67, median: 33
5) mean: 18.8, median: 15
6) mean: 22.5, median: 15
7) mean: 29.4, median: 22.5
8) mean: 24, median: 22
9) mean: 41.4, median: 32
10) mean: 12, median: 11
11) mean: 17.5, median: 15
12) mean: 20, median: 16
13) mean: 9.86, median: 6
14) mean: 22.86, median: 22
15) mean: 22, median: 23
16) mean: 17.43, median: 16
17) mean: 15.63, median: 14.5
18) mean: 37.5, median: 37
19) mean: 28, median: 28.5
20) mean: 35.71, median: 37

Mode and Range

1) mode: 16, range: 63
2) mode: 18, range: 14
3) mode: 25, range: 36
4) mode: 14, range: 34
5) mode: 18, range: 44
6) mode: 12, range: 16
7) mode: 34, range: 41
8) mode: 72, range: 65
9) mode: 18, range: 34
10) mode: 1, range: 18
11) mode: 9, range: 13
12) mode: 21, range: 34
13) mode: 5, range: 27
14) mode: 14, range: 17
15) mode: 16, range: 28
16) mode: 8, range: 40
17) mode: 19, range: 32
18) mode: 48, range: 61
19) mode: 15, range: 69
20) mode: 67, range: 53

Stem–And–Leaf Plot

1)

Stem	leaf
2	2 5 6 8 9
4	0 1 3 6 7 8
6	5 6

2)

Stem	leaf
1	0 9
3	0 2 5 6 8 9
5	1 2 4 9

3)

Stem	leaf
4	1 5 5 6 9
8	5 8 9
11	2 4 6 7

4)

Stem	leaf
5	1 9 5
7	0 1 2 2 6 9
10	1 3 6

5)

Stem	leaf
2	7
6	3 4 8
13	3 7 9

6)

Stem	leaf
3	2 7 7
9	0 2 5 7
14	1 1 6 8

Dot plots

1) 23
2) 5
3) 10

4) 6 and 12
5) 2
6) 4

7) 2

Bar Graph

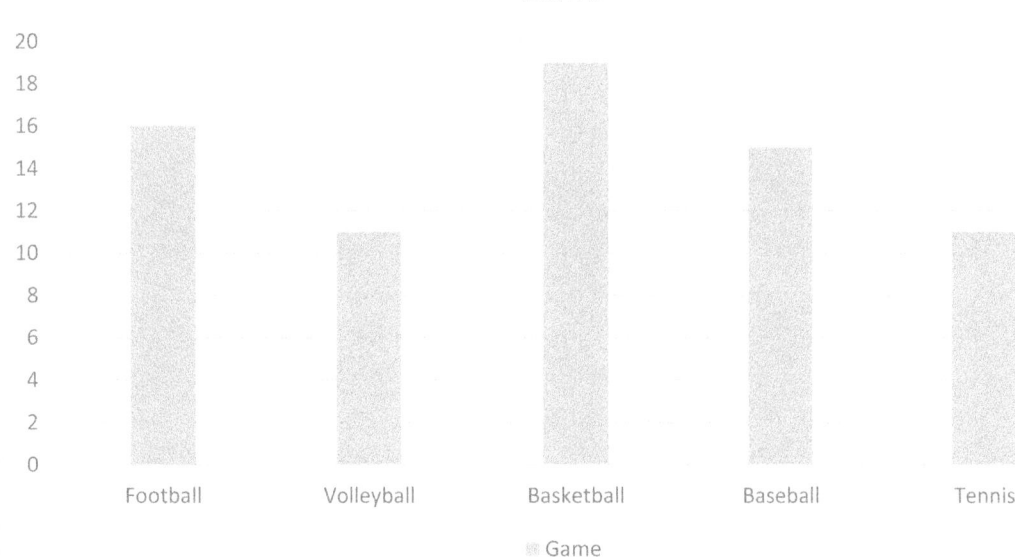

1) Basketball
2) 8 students

3) Volleyball and Tennis
4) 27

5) Football
6) Volleyball and Tennis

Probability

1) $\frac{1}{5}$
2) $\frac{5}{6}$
3) $\frac{1}{6}$

4) $\frac{1}{3}$
5) $\frac{1}{10}$
6) $\frac{5}{12}$

7) $\frac{5}{13}$
8) $\frac{1}{8}$

SBAC Test Review

SBAC GRADE 5 MAHEMATICS REFRENCE MATERIALS

Perimeter

Square $\qquad$ $P = 4S$

Rectangle $\qquad$ $P = 2L + 2W$

Area

Square $\qquad$ $A = S \times S$

Rectangle $\qquad$ $A = l \times w$ or $A = bh$

Volume

Cube $\qquad$ $A = S \times S \times S$

Rectangular Prism $\qquad$ $A = l \times w \times h$ or $A = Bh$

LENGTH

Customary	Metric
1 mile (mi) = 1,760 yards (yd)	1 kilometer (km) = 1,000 meters (m)
1 yard (yd) = 3 feet (ft)	1 meter (m) = 100 centimeters (cm)
1 foot (ft) = 12 inches (in.)	1 centimeter (cm) = 10 millimeters (mm)

VOLUME AND CAPACITY

Customary	Metric
1 gallon (gal) = 4 quarts (qt)	1 liter (L) = 1,000 milliliters (mL)
1 quart (qt) = 2 pints (pt.)	
1 pint (pt.) = 2 cups (c)	
1 cup (c) = 8 fluid ounces (Fl oz)	

WEIGHT AND MASS

Customary	Metric
1 ton (T) = 2,000 pounds (lb.)	1 kilogram (kg) = 1,000 grams (g)
1 pound (lb.) = 16 ounces (oz)	1 gram (g) = 1,000 milligrams (mg)

Smarter Balanced Assessment Consortium

SBAC Practice Test 1

Mathematics

GRADE 5

- ❖ 30 Questions
- ❖ There is no time limit for this practice test.
- ❖ Calculators are NOT permitted for this practice test

Administered *Month Year*

SBAC Math Practice Grade 5

1) Which expression correctly shows the sum of product of 6 and 14 and the difference of 17 and 5?

   A. $14 + (6 \times 17) - 5$

   B. $(14 \times 6) + (17 - 5)$

   C. $(14 \times 6) - (17 - 5)$

   D. $14 - (6 \times 17) + 5$

2) Which equation has the same unknown value as $390 \div 26 = \square$?

   A. $390 \times \square = 26$

   B. $\square \div 390 = 26$

   C. $26 \times \square = 390$

   D. $\square \div 26 = 390$

3) The owner of a snow-cone stands used $\frac{1}{8}$ gallon of syrup to make 16 cherry snow cones. She used the same amount of syrup in each snow cone. How much syrup in gallons was used in each cherry snow cone?

   A. $\frac{1}{2}$ gal

   B. 2 gal

   C. $\frac{1}{128}$ gal

   D. 128 gal

4) 29 students equally share a bag of 522 dimes and 720 nickels. How many dimes does each student get?

   A. 18

   B. 22

   C. 32

   D. 98

5) Bertha bought 7 cans of tuna at $1.32 a can. How much did she spend?

   A. $6.20

   B. $9.24

   C. $9.40

   D. $19.42

6) Which equation shows how to multiply $13 \times 8 \times 6$ using the associative property?

   A. $13 \times 8 \times 6 = 6 \times 9 \times 13$

   B. $(13 \times 8) + (13 \times 6) = (13 + 6) \times (13 + 8)$

   C. $(13 \times 8) + 6 = 13 \times (8 + 6)$

   D. $(13 \times 8) \times 6 = 13 \times (8 \times 6)$

7) What is 0.73 rounded to the tenths place?

   A. 0.6

   B. 0.7

   C. 0.8

   D. 1.0

8) Each time Elijah goes to the movies he spends $13.00. Which expression shows how much he spends after going to the movies t times?

A. $13.00 + t

B. $13.00 − t

C. $13.00 ÷ t

D. $13.00 × t

9) Benjamin is using a calculator to multiply 5,649 and 30. He enters 5,649 × 300 by mistake. What can Benjamin do to correct his mistake?

A. Subtract 330 from the product

B. Add 330 to product

C. Divide the product by 10

D. Multiply the product by 10

10) The price of shoes in a store is $30 and the price of belt in the same store is $8. A customer buys 4 shoes and 5 belts during a sale when the price of shoes is discounted 35% and the price of belt is discounted 6%. How much did the customer save due to the sale?

A. $44.4

B. $4.44

C. $42.4

D. $40.44

11) Daniel wrote the expression shown. $28 \div 7 + 14(57 - 8)$. What do these parentheses indicate in the expression?

   A. Divide 28 by 7 before adding 14

   B. Multiply 14 by 57 before subtracting 8

   C. Subtract 8 from 57 before multiplying by 14

   D. Add 7 and 14 together before subtracting 8 from 57

12) Lucas is doing his homework. It takes him 14 minutes to do 6 problems of math and 18 minutes to read 4 pages of biology. Lucas reads the same amount of biology pages each minute. Which of these is closest to the number of minutes it takes Lucas to read each page of his biology homework?

   A. 4 minutes

   B. $4\frac{1}{2}$ minutes

   C. 2 minutes

   D. $4\frac{1}{2}$ minutes

13) What is the sum of angles α + β in the right triangle below?

   A. 60 degrees

   B. 45 degrees

   C. 90 degrees

   D. 360 degrees

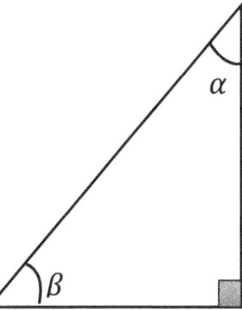

14) Which statement describe $\frac{5}{6} \times \frac{3}{8}$?

A. $\frac{5}{6} \times \frac{3}{8}$ is 5 group of $\frac{3}{8}$, divided into 48 equal parts.

B. $\frac{5}{6} \times \frac{3}{8}$ is 5 group of $\frac{3}{8}$, divided into 6 equal parts.

C. $\frac{5}{6} \times \frac{3}{8}$ is 6 group of $\frac{3}{8}$, divided into 5 equal parts.

D. $\frac{5}{6} \times \frac{3}{8}$ is 6 group of $\frac{3}{8}$, divided into 15 equal parts.

15) Noah and Liam played a new game for 10.5 hours last week. If they played the same amount of time each of 7 days, how long did they play each day?

A. 0.50 hour

B. 0.75 hour

C. 1.05 hour

D. 1.50 hour

16) Daniel planted five cucumber seeds. Out of the five planted, only three sprouted. How many plants can Daniel plan on yielding if he plants 200 seeds?

Number of seeds	7				420
Successes	4				?

A. 80

B. 40

C. 160

D. 240

17) A wire is 24 feet long. The Robert needs to cut pieces that are $\frac{6}{7}$ foot long. How many pieces can he cut?

   A. 28

   B. 18

   C. 14

   D. 7

18) The temperature was 48°F at 9:00 A.M., 53°F at 9:30 A.M., and 58°F at 10:30 A.M. Describe the pattern, and predict the temperature at 12:30 A.M.

   A. Add 6°F; 87°F

   B. Add 5°F; 87°F

   C. Add 6°F; 78°F

   D. Add 5°F; 78°F

19) A rectangular prism measures 8-unit cubes wide and 5-unit cubes high. If the volume of the prism is 160 cubic unit, what is the length of the prism in unit cubes?

   A. 4

   B. 15

   C. 14

   D. 24

SBAC Math Practice Grade 5

20) Charlotte has a grosgrain ribbon 7 feet long. She cuts the ribbon into 9 equal pieces. Which equation shows how to find the length, in feet of each piece of the ribbon?

   A. $9 \times 7 = 40$

   B. $7 \div 9 = \frac{7}{9}$

   C. $9 + 7 = 16$

   D. $9 \div 7 = 1\frac{2}{7}$

21) Paul and his three friends had lunch together. The total bill for lunch came to $29.80, including tip. If they shared the bill equally, how much did they each pay?

   A. 7.15

   B. 6.45

   C. $7.45

   D. 10.15

22) What is the volume of the figure in cubic units?

   A. 58

   B. 147

   C. 127

   D. 87

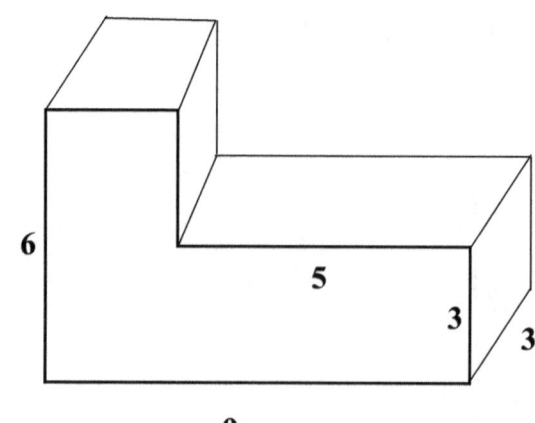

23) Eugen has a piggy bank filled with coins.

- 16 are quarters
- 17 are dimes
- 23 are nickels
- 44 are pennies

What percent of the coin are nickels?

A. 23%

B. 77%

C. 67%

D. 100%

24) Sophia is buying crullers for a group of 30 students. If 70% of the group want crullers, and each student will eat 3 crullers, how many crullers should Sophia buy?

A. 6 crullers

B. 21 crullers

C. 63 crullers

D. 84 crullers

25) Which fraction is equivalent to $\frac{5}{8}$?

A. $\frac{25}{56}$

B. $\frac{5}{4}$

C. $\frac{20}{48}$

D. $\frac{20}{32}$

SBAC Math Practice Grade 5

26) Jayden is having a party. He wants to get three cookies for each of the 4 people, including himself, who will be at the party. If each cookie costs 60 ₵, how much money will he spend on cookies? which diagram below best shows this problem?

A.   B. 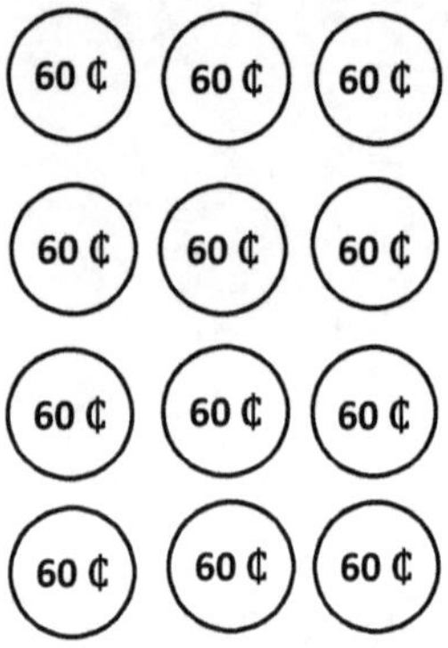   C. 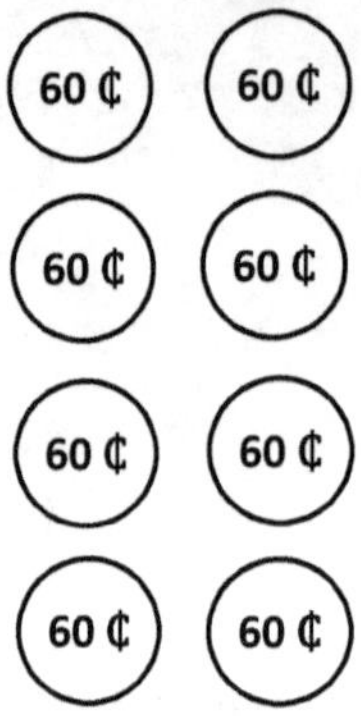   D.

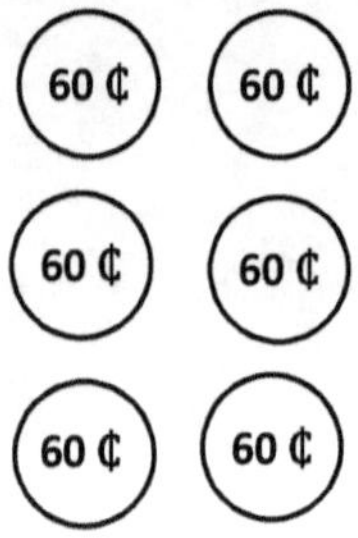

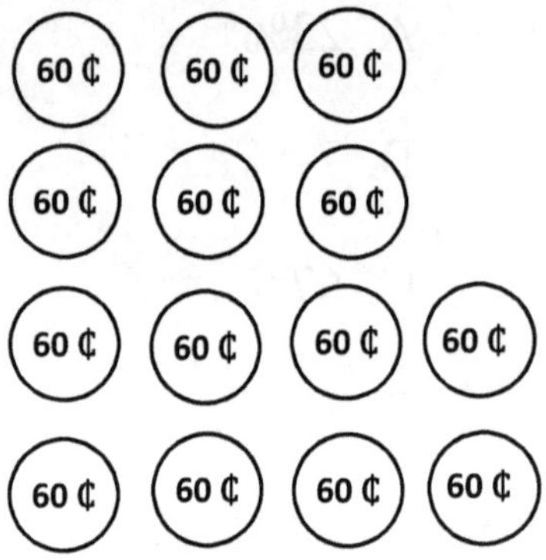

27) What is the perimeter of the rectangular in the figure below?

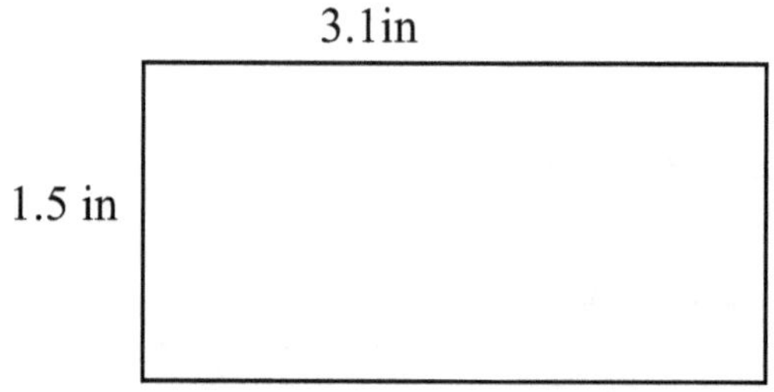

A. 4.6

B. 4.65

C. 9

D. 9.2

28) What is the value of C in the table below?

A. 64

B. 129

C. 150

D. 136

Input	Output
2	3
6	15
14	39
30	87
51	C

29) Robert went fishing and caught 3 fish. They weighed 16.86 ounces, 34.98 ounces, and 112.25 ounces. How much did all 3 Robert's fishes weigh together?

A. 96.05 ounces

B. 146.50 ounces

C. 164.09 ounces

D. 246.05 ounces

30) The bar graph shows the monthly high sale for a grocery store in 2018. According to the graph, for all months shown, how much smaller is the mean than the median sale?

A. 0 Million Dollars

B. 30 Million Dollars

C. 5 Million Dollars

D. 15 Million Dollars

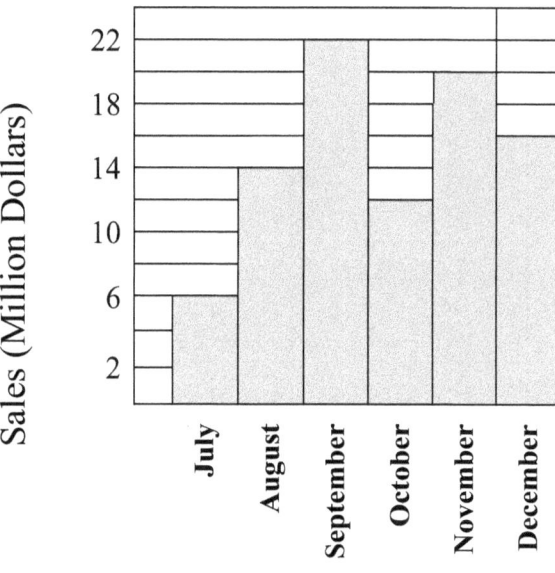

Smarter Balanced Assessment Consortium
SBAC Practice Test 2

Mathematics
GRADE 5

- ❖ 30 Questions
- ❖ There is no time limit for this practice test.
- ❖ Calculators are NOT permitted for this practice test

Administered *Month Year*

SBAC Math Practice Grade 5

1) Which digit is the thousands digit in the number 34,120,097?

    A. 1

    B. 2

    C. 0

    D. 3

2) What is the value of expression shown? $3[7.5 - 3(1.2)]$

    A. 3.6

    B. 1.6

    C. 16

    D. 13.6

3) Which expression equal to $\frac{63}{147}$?

    A. $\frac{4}{7}$

    B. $\frac{3}{8}$

    C. $\frac{2}{7}$

    D. $\frac{3}{7}$

4) How can the distributive property be used to solve this expression? 48×19

    A. $(40 + 10) \times (8 \div 9)$

    B. $(4 \times 1) + (8 \times 9)$

    C. $(48 + 9) \times (48 + 1)$

    D. $(48 \times 10) + (48 \times 9)$

5) Which statement correctly compares the two values?

   A. The value of 7 in 9.73 is 10 times the value of the 7 in 7.39

   B. The value of 7 in 9.73 is $\frac{1}{10}$ the value of the 7 in 7.39

   C. The value of 7 in 9.73 is 100 times the value of the 7 in 7.39

   D. The value of 6 in 9.73 is $\frac{1}{100}$ the value of the 7 in 7.39

6) What is 75.28 ÷ 100?

   A. 7.528

   B. 0.7528

   C. 75.28

   D. 752.8

7) Which number represents six million fifty thousand four hundred four?

   A. 6,050,044

   B. 6,500,404

   C. 6,050,404

   D. 6,500,044

8) Betty has 0.8 liters of juice. How many milliliters(mL) of juice does Betty have?

   A. 0.008 mL

   B. 0.0008 mL

   C. 8,000 mL

   D. 800 mL

SBAC Math Practice Grade 5

9) Mr. Martinez had 16 daylily plants. Each plant produced 728 flowers. How many flowers did the plants produce?

   A. 86

   B. 868

   C. 1,648

   D. 11,648

10) The table below shows the lengths of different Stamps on display at a post office. Which stamps has the shortest length?

   A. Stamp 1

   B. Stamp 2

   C. Stamp 3

   D. Stamp 4

Stamp	Length (cm)
1	$\frac{1}{6}$
2	$\frac{3}{4}$
3	$\frac{3}{8}$
4	$\frac{1}{2}$

11) Find the value of the nth term in the sequence.

Position	9	12	15	21	n
Value of Terms	2	3	4	6	

   A. $\frac{n}{4}$

   B. $n-7$

   C. $\frac{n}{3} - 1$

   D. $\frac{n}{3} + 1$

WWW.MathNotion.com

12) The dot plot shows the number of students who made from 1 to 8 words. What fraction of the student in the class made 6 or more words?

A. $\frac{1}{6}$

B. $\frac{1}{3}$

C. $\frac{3}{4}$

D. $\frac{1}{4}$

13) Jessica flips a coin 80 times. The coin lands on head 24 times. What percent of the flips were tails?

A. 20%

B. 80%

C. 30%

D. 70%

14) A parallelogram has a base of 31 cm and the height of 28 cm. a triangle has a base of 58 cm and a height of 31 cm. which figure has the greater area? By how much? (*Area of parallelogram = base × high*)

A. Parallelogram, 31 cm^2

B. Parallelogram, 3 cm^2

C. Triangle, 31 cm^2

D. Triangle, 3 cm^2

15) Which of the following fractions is higher than $\frac{1}{8}$, but less than $\frac{1}{3}$?

   A. $\frac{8}{9}$

   B. $\frac{3}{4}$

   C. $\frac{1}{2}$

   D. $\frac{1}{4}$

16) $\frac{43}{50}$ may be written as a percent as:

   A. 43%

   B. 16%

   C. 24%

   D. 86%

17) A chart below is showing the socks that Jenna has in her dresser. What is the probability that Jenna will choose a red sock at random?

   A. $\frac{1}{7}$

   B. $\frac{1}{3}$

   C. $\frac{1}{21}$

   D. $\frac{1}{14}$

Color	Number
Purple	2
White	3
Red	7
Stripped	5
Black	4

18) The measures of three of the interior angles of a quadrilateral are 70°, 100°, and 95°. What is the measure of the fourth angle of this quadrilateral?

   A. 90°

   B. 95°

   C. 100°

   D. 125°

19) What is the greatest common factor (GCF) of 63 and 36?

   A. 4

   B. 6

   C. 9

   D. 18

20) Which inequality is true?

   A. 0.23 > 0.32

   B. 0.98 < 0.89

   C. 0.43 > 0.44

   D. 0.56 < 0.65

21) Which percent is equivalent to 0.6?

   A. 6%

   B. 0.6%

   C. 0.006%

   D. 60%

SBAC Math Practice Grade 5

22) James paid $14.70 each for 4 used video games and $26.4 each for 3 new video games. How much money did James spend on video games?

   A. $138

   B. $105.06

   C. $95.60

   D. $95.06

23) The arcade changes the price shown below for game tokens. Using the pattern shown, what is the price for 5 tokens?

   A. $2.00

   B. $4.00

   C. $40.00

   D. $20.00

Tokens	Amount
10	$4.00
20	$8.00
30	$12.00
40	$16.00
50	$20.00

24) Suzy's class used $\frac{7}{9}$ of the cafeteria trays for their science display. There are 63 trays in all. How many trays were used?

   A. 7

   B. 81

   C. 49

   D. 270

25) The perimeter of the trapezoid below is 36. What is its area?

A. 46 cm

B. 36 cm

C. 30 cm

D. 46.5 cm

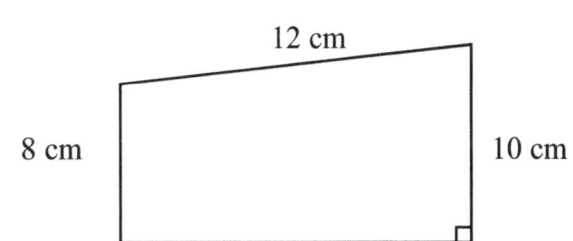

26) Joseph's first planter is 8 feet long and 3 feet wide and 4 feet height. The container is filled with soil to a height of 3 feet in the planter. What is the volume of soil in the planter?

A. 80 cubic cm

B. 116 cubic cm

C. 32 cubic cm

D. 96 cubic cm

27) The stem and leaf plot shows the numbers of miles run by each of runners. What is the difference between the number of runners who ran fewer than 40 miles and the number of runners who ran more than 75 miles?

A. 14

B. 15

C. 3

D. 2

Stem	Leaf
0	7 8
1	0
2	0 9
3	2 7 7 9
6	4 4 8
8	2 9
9	5 5 6 9

SBAC Math Practice Grade 5

28) Linda will order ice-cream for her party. The relationship between S, the number of the scoops she will order, and C, the total costs of ice-cream she will pay, can be represented by the equation $C = 7S$.

Which table contain only values that represent the equation?

A. **Invoice**

Number of Scoops (S)	Cost (C)
4	$28
9	$72
11	$77
13	$91

B. **Invoice**

Number of Scoops (S)	Cost (C)
4	$28
9	$63
11	$77
13	$91

C. **Invoice**

Number of Scoops (S)	Cost (C)
1	$7
6	$42
9	$72
16	$112

D. **Invoice**

Number of Scoops (S)	Cost (C)
1	$7
6	$42
9	$72
16	$108

29) A cube has a volume of 27 cubic units. How many unit cubes will fit along one side of the cube if there are no gaps on overlaps?

A. 3

B. 9

C. 14

D. 54

30) Figure M and figure N are congruent. What is the length of side b in figure N?

A. 3

B. 9

C. 6

D. 8

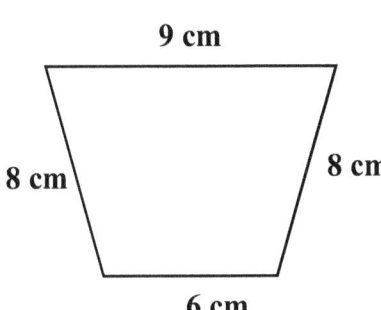
Figure M

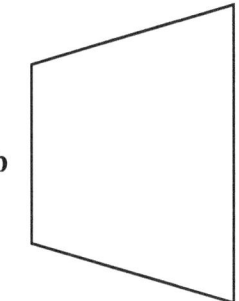
Figure N

SBAC Math Practice Grade 5

Answers and Explanations

Answer Key

Now, it's time to review your results to see where you went wrong and what areas you need to improve!

SBAC Math Practice Tests

Practice Test 1

1	B	11	C	21	C
2	C	12	B	22	B
3	C	13	C	23	A
4	A	14	B	24	C
5	B	15	D	25	D
6	D	16	D	26	B
7	B	17	A	27	D
8	D	18	D	28	C
9	C	19	A	29	C
10	A	20	B	30	A

Practice Test 2

1	C	11	C	21	D
2	A	12	B	22	A
3	D	13	D	23	A
4	D	14	C	24	C
5	A	15	D	25	B
6	B	16	D	26	D
7	C	17	B	27	C
8	D	18	B	28	B
9	D	19	C	29	A
10	A	20	D	30	C

Practice Test 1

Answers and Explanations

1) Answer: B

The product of 6 and 14 is (6×14)

The difference of 17 and 5: $(17 - 5)$

Then, the sum: $(6 \times 14) + (17 - 5)$

2) Answer: C

Multiplication is the inverse of division and you can use multiplication to check your division answer (Dividend÷Divisor =Quotient; Then, Quotient × Divisor = Dividend).

$390 \div 26 = \Box \rightarrow \Box \times 26 = 390$

3) Answer: C

$\frac{1}{8} \div 16 = \frac{1}{8} \times \frac{1}{16} = \frac{1}{128}$

4) Answer: A

The bag of dimes should be divided by all students: $522 \div 29 = 18$ dimes

5) Answer: B

$7 \times 1.32 = \$9.24$

6) Answer: D

Associative property of multiplication: when multiplying three or more real numbers, the product is always the same regardless of their regrouping. By 'grouping' we mean how you use parenthesis. (a×b) ×c=a×(b×c)

7) Answer: B

To round a number to the nearest tenth, look at the next place value to the right (the hundredths). If it is 4 or less, just remove all the digits to the right. If it is 5 or greater, add 1 to the digit in the tenths place, and then remove all the digits to the right.

8) Answer: D

Each time spend $13 and, t times, means multiplication, then we need to use multiplication. The answer is: 13t

SBAC Math Practice Grade 5

9) Answer: C

$5{,}649 \times 300 = 5{,}649 \times 30 \times 10$, then $\frac{5{,}649 \times 30 \times 10}{10} = 5{,}649 \times 30$

10) Answer: A

4 Shoes: $4 \times 30 = 120$; 35% of 120: $0.35 \times 120 = \$42$

5 Belts: $5 \times 8 = 40$; 6% of 40: $0.06 \times 40 = \$2.4$

$\$42 + \$2.4 = \$44.4$ the customer saved

11) Answer: C

The **PEMDAS Rule** (an acronym for "Parenthesis, Exponents, Multiplication, Division, Addition, Subtraction") is a set of rules that prioritize the order of calculations, that is, which operation to perform first. Parenthesis in math are used to group important things together, so you always do them first.

12) Answer: B

$18 \div 4 = \frac{18}{4} = 4\frac{2}{4} = 4\frac{1}{2}$ minutes

13) Answer: C

The sum of the measure of the three interior angles, in every triangle, is $180°$.

A right triangle is a type of triangle that has one angle that measures $90°$, then:

$90° + \alpha + \beta = 180° \rightarrow \alpha + \beta = 180° - 90° = 90°$

14) Answer: B

To multiply fractions is following the 3 steps: Multiply the numerators, Multiply the denominators, and Simplify the resulting fraction.

$\frac{5}{6} \times \frac{3}{8} = (\frac{1}{6} \times \frac{5}{1}) \times \frac{3}{8} = \frac{1}{6} \times (5 \times \frac{3}{8}) = \frac{5 \times \frac{3}{8}}{6}$

15) Answer: D

$10.5 \div 7 = 1.5$

16) Answer: D

Equivalent Fractions have the same value, even though they may look different. Because when you multiply or divide both the top and bottom by the same number, the fraction keeps its value. Then, $\frac{7}{4} = \frac{420}{?} \rightarrow \frac{7 \times 60}{4 \times 60} = \frac{420}{240}$

SBAC Math Practice Grade 5

17) Answer: A

24 feet long wire is cut into $\frac{6}{7}$ equal parts. Therefore, 24 should be divided by $\frac{6}{7}$

$24 \div \frac{6}{7} = 24 \times \frac{7}{6} = 28$

18) Answer: D

The pattern is: 48°, 53°, 58° (Add 5°).

12:30A.M.−10:30A.M.=2:00 hours, and change to the half hours is $2 \times 2 = 4$, and use the pattern: $4 \times 5 = 20$; then, $20° + 58° = 78°$

19) Answer: A

Use volume of rectangle prism. V = width × length × heigth

$\Rightarrow 160 = 8 \times L \times 5 \Rightarrow 160 = 40L \Rightarrow L = \frac{160}{40} = 4$ uint-cubes

20) Answer: B

7 feet long ribbon is cut into 9 equal parts. Therefore, 7 should be divided by 9

21) Answer: C

$29.80 \div 4 = \$7.45$

22) Answer: B

We have two rectangular prisms.

Use volume of rectangle prism. V = width × length × heigth

$V_1 = 5 \times 3 \times 5 = 75$ cubic units

$V_2 = 6 \times 4 \times 3 = 72$ cubic units

$V = V_1 + V_2 = 75 + 72 = 147$ cubic unit

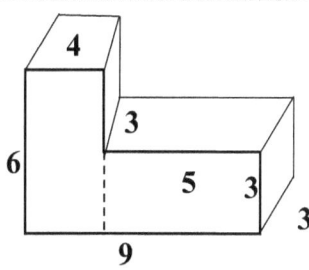

23) Answer: A

Percent means "out of 100." Or, Use percent formula: percent $= \frac{part}{whole} \times 100$

Whole= $16 + 17 + 23 + 44 = 100$,

The nickels are 23, means: $\frac{23}{100}$ or 23%

24) Answer: C

Use percent formula: part $= \frac{percent}{100} \times$ whole

WWW.MathNotion.com

Part = $\frac{70}{100} \times 30 = \frac{30 \times 70}{100} = \frac{2,100}{100} = 21$ students,

$21 \times 3 = 63$ crullers

25) Answer: D

Multiplying or dividing the numerator and denominator of a fraction by the same number will produce an equivalent fraction. $\frac{5}{8} \times \frac{4}{4} = \frac{20}{32}$

26) Answer: B

He wants to buy 3 cookies for 4 people: $3 \times 4 = 12$ cookies

Each cookie cost 60 cents, then, we need to have 12 of ₡60 or 12×60.

27) Answer: D

Perimeter is P = 2l + 2w, where l is the length and w is the width

$p = 2(3.1) + 2(1.5) = 6.2 + 3 = 9.2$ in.

28) Answer: C

We need to find the relation between input and output.

Input: $(6 - 2) = 4$, Output: $(15 - 3) = 12$, and the relation is 3 times of 4 is 12

Input, $(51 - 30) = 21$, use the relation: $3 \times 21 = 63$, then:

Output, $C - 87 = 63 \rightarrow C = 63 + 87 = 150$

29) Answer: C

$16.86 + 34.98 + 112.25 = 164.09$

30) Answer: A

The sales are: 6, 14, 22, 12, 20, and 16

Write the numbers in order: 6, 12, 14, 16, 20, 22

Median: $\frac{14+16}{2} = \frac{30}{2} = 15$

Mean: $\frac{\text{sum of terms}}{\text{number of terms}} = \frac{6+12+14+16+20+22}{6} = \frac{90}{6} = 15$

Median − Mean = 15 − 15 = 0

Practice Test 2

Answers and Explanations

1) Answer: C

Place Value Chart								
Millions			Thousands			Ones		
Hundreds	Tens	Ones	Hundreds	Tens	Ones	Hundreds	Tens	Ones
	3	4	1	2	0	0	9	7

In the number 34,120,097 the digit 0 is in the (one) thousands place.

2) Answer: A

You should have used the order of operations, or PEMDAS. 1. Operations contained in Parentheses or brackets, 2. Exponents 3. Multiplication/Division from left to right, and 4. Addition/Subtraction from left to right.

$3[7.5 - 3(2.1)] = 3[7.5 - 6.3] = 3[1.2] = 3.6$

3) Answer: D

Multiplying or dividing the numerator and denominator of a fraction by the same number will produce an equivalent fraction. $\frac{63}{147}$ divided by 7 = $\frac{9}{21}$ divided by 3 = $\frac{3}{7}$

4) Answer: D

The distributive property lets you multiply a sum by multiplying each addend separately and then add the products.

$48 \times 19 = 48 \times (10 + 9) = (48 \times 10) + (48 \times 9)$

5) Answer: A

The place value of 7 in 9.73 is tenth

The place value of 7 in 7.39 is ones

To convert tenth place to ones we need to multiply by 10.

6) Answer: B

$75.28 \div 100 = 0.7528$

SBAC Math Practice Grade 5

7) Answer: C

When converting word names to standard form. The words "million" and "thousand" tell you which periods the digits are in.

8) Answer: D

1 L = 1,000 mL; Convert 0.8 L to mL: $0.8 \times 1,000 = 800$ mL

9) Answer: D

$16 \times 728 = 11,648$

10) Answer: A

Compare the fractions: $\frac{3}{4} > \frac{1}{2} > \frac{3}{8} > \frac{1}{6}$

Therefore, $\frac{1}{6}$ is the least fraction and related to stamp 1

11) Answer: C

Plug in $n = 9$ and $n = 12$ in each equation.

A. $\frac{n}{4} \rightarrow \frac{9}{4} \neq 2$

B. $n - 7 \rightarrow 9 - 7 = 2$ and $12 - 7 = 5 \neq 3$

C. $\frac{n}{3} - 1 \rightarrow \frac{9}{3} - 1 = 3 - 1 = 2$ and $\frac{12}{3} - 1 = 4 - 1 = 3$ Bingo!

D. $\frac{n}{3} + 1 \rightarrow \frac{9}{3} + 1 = 3 + 1 \neq 2$

12) Answer: B

All student is 12, and 4 of them made 6 and more, then $\frac{4}{12}$ and simplify: $\frac{1}{3}$

13) Answer: D

Use percent formula: $percent = \frac{part}{whole} \times 100$

$Percent = \frac{24}{80} \times 100 = 30\%$ coin lands on head. 100%−30%=70% land on tail.

14) Answer: C

Area of parallelogram: $A = b.h = 31 \times 28 = 868 \ cm^2$

Area of triangle: $A = \frac{1}{2} b.h = \frac{1}{2} \times 58 \times 31 = 899 \ cm^2$

Area of triangle− Area of parallelogram $= 899 - 868 = 31 \ cm^2$

SBAC Math Practice Grade 5

15) Answer: D

Compare the fractions:

$\frac{8}{9} > \frac{1}{3}$; $\frac{3}{4} > \frac{1}{3}$; $\frac{1}{3} < \frac{1}{2}$; $\frac{1}{4} > \frac{1}{8}$, and $\frac{1}{4} < \frac{1}{3}$

16) Answer: D

Convert fraction to decimal: $\frac{43}{50} \times 100 = \frac{43 \times 100}{50}$ both numerator and denominator divided by 50: $\frac{43 \times 2}{1} = 86\%$

17) Answer: B

Probability $= \frac{number\ of\ desired\ outcomes}{number\ of\ total\ outcomes} = \frac{7}{2+3+7+5+4} = \frac{7}{21} = \frac{1}{3}$

18) Answer: B

Quadrilaterals are four sided polygons, with four vertexes, whose total interior angles add up to 360 degrees.

$70° + 100° + 95° = 265°$; $360° - 265° = 95°$

19) Answer: C

Factor of 63: $(1, 3, 7, 9, 21, 63)$

Factor of 36: $(1, 2, 3, 4, 6, 9, 12, 18, 36)$

Greatest Common Factor is: 9

20) Answer: D

when decimals are compared start with tenths place and then hundredths place, etc. If one decimal has a higher number in the tenths place, then it is larger than a decimal with fewer tenths. If the tenths are equal compare the hundredths, then the thousandths etc. $0.56 < 0.65$ is correct.

21) Answer: D

Convert decimal to percent by multiply 100: $0.6 \times 100 = 60\%$

22) Answer: A

$4 \times \$14.70 = \$58.80$, and $3 \times \$26.4 = \$79.2$

$\$58.80 + \$79.2 = \$138$

23) Answer: A

The rule is $y = 0.40\ x$, when y is output (amount) and x is input (Token).

Amount of 5 token is: $Amount = 0.4 \times 5 = \$2.00$

24) Answer: C

$\frac{7}{9} \times 63 = \frac{7 \times 63}{9} = \frac{7 \times \cancel{63}^{7}}{\cancel{9}_{1}} = 49$

25) Answer: B

First, find the missing side of the trapezoid. The perimeter of the trapezoid below is 36.

Therefore, the missing side of the trapezoid (its height) is:

$36 - (8 + 10 + 12) = 36 - 30 = 6$

Area of a trapezoid: $A = \frac{1}{2} h (b1 + b2) = \frac{1}{2} (6)(10 + 8) = 54$ cm

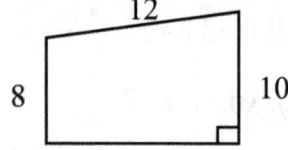

26) Answer: D

Use volume of rectangle prism formula.

$V = length \times width \times height \Rightarrow V = 8 \times 3 \times 4 = 96$ cubic cm

27) Answer: C

There are 9 values on the stem and leaf plot that are less than 40 (7, 8, 10, 20, 32, 37, 37, and 39) and 6 values on the stem and leaf plot that are greater than 75 (82, 89, 95, 95, 96 and 99). Then subtracted 6 from 9, resulting in a difference of 3.

28) Answer: B

To determine each C-value in the table, we should have multiplied 7 by each S-value: ($4 \times 7 = 28$; $9 \times 7 = 63$; $11 \times 7 = 77$; and $13 \times 7 = 91$).

29) Answer: A

Cube is a three-dimensional shape that has equal width, height, and length measurements. If a cube has side length "S" then volume is: $V = S \times S \times S = S^3$

$V = S \times S \times S \rightarrow 27 = S \times S \times S$, plug in the small number of options provided to find the length of side. If $S = 3$ $V = 3 \times 3 \times 3 = 27$, means $S = 3$ cubes

30) Answer: C

When shapes are congruent, all corresponding sides and angles are also congruent. The same shape and size, but we can flip, slide or turn. Then, $b = 6 cm$

"End"

www.ingramcontent.com/pod-product-compliance
Lightning Source LLC
LaVergne TN
LVHW061310060426
835507LV00019B/2085